Identi

PIERID

11

12

SATYRIDAE

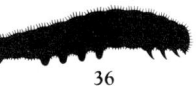

36

LASIOCAMPIDAE

42

...IIDAE

47

ENDROMIIDAE

48

...AE GEOMETRIDAE

73

GEOMETRIDAE

79

The Observer's Pocket Series

CATERPILLARS

Observer's Books

NATURAL HISTORY
Birds · Birds' Eggs · Wild Animals · Zoo Animals
Farm Animals · Freshwater Fishes · Sea Fishes
Tropical Fishes · Butterflies · Larger Moths · Caterpillars
Insects · Pond Life · Sea and Seashore · Seashells · Pets
Dogs · Horses and Ponies · Cats · Trees · Wild Flowers
Grasses · Mushrooms · Lichens · Cacti · Garden Flowers
Flowering Shrubs · Vegetables · House Plants · Geology
Rocks and Minerals · Fossils · Weather · Astronomy

SPORT
Soccer · Cricket · Golf · Tennis · Coarse Fishing
Fly Fishing · Show Jumping · Motor Sport

TRANSPORT
Automobiles · Aircraft · Commercial Vehicles
Motorcycles · Steam Locomotives · Ships · Small Craft
Manned Spaceflight · Unmanned Spaceflight

ARCHITECTURE
Architecture · Churches · Cathedrals · Castles

COLLECTING
Awards and Medals · Coins · Firearms · Furniture
Postage Stamps · Glass · Pottery and Porcelain

ARTS AND CRAFTS
Music · Painting · Modern Art · Sewing · Jazz
Big Bands

HISTORY AND GENERAL INTEREST
Ancient Britain · Flags · Heraldry · European Costume

TRAVEL
London · Tourist Atlas GB · Lake District · Cotswolds

The Observer's Book of
CATERPILLARS

DAVID J. CARTER

WITH 226 COLOUR ILLUSTRATIONS
BY GORDON RILEY
AND 17 BLACK AND WHITE DRAWINGS

FREDERICK WARNE

LONDON

First published 1979 by
Frederick Warne (Publishers) Ltd, London
© 1979 Frederick Warne (Publishers) Ltd

ISBN 0 7232 1592 8

Printed in Great Britain by
Butler & Tanner Ltd, Frome and London
0080·1178

Contents

List of Colour Plates	7
Introduction	11
Notes	39
Butterflies	41
Moths	65
Bibliography	152
Index	153

Note: For page references to individual caterpillars please see Index.

List of Colour Plates

Note: Where the caterpillars shown in the colour plates are not of average size, this is indicated on the relevant plate. The numbers refer to text entries.

Plate 1   1 Small Skipper, 2 Large Skipper, 3 Dingy Skipper, 4 Grizzled Skipper, 5 Swallow-tail, 5a Swallow-tail with gland everted

Plate 2   6 Clouded Yellow, 7 Brimstone, 8 Large White, 9 Small White, 10 Green-veined White, 11 Orange Tip

Plate 3   12 Green Hairstreak, 13 Purple Hairstreak, 14 Small Copper, 15 Small Blue, 16 Common Blue, 17 Chalkhill Blue, 18 Holly Blue

Plate 4   19 White Admiral, 20 Purple Emperor, 21 Red Admiral, 22 Painted Lady, 23 Small Tortoiseshell, 24 Peacock

Plate 5   25 Comma, 26 Small Pearl-bordered Fritillary, 27 Pearl-bordered Fritillary, 28 High Brown Fritillary, 29 Dark Green Fritillary, 30 Silver-washed Fritillary

Plate 6   31 Speckled Wood, 32 Wall, 33 Marbled White, 34 Grayling, 35 Gatekeeper, 36 Meadow Brown, 37 Small Heath, 38 Ringlet

Plate 7   39 December Moth, 40 Pale Oak Eggar, 41 Small Eggar, 42 Lackey, 43 Oak Eggar, 44 Fox Moth

Plate 8   45 Drinker, 46 Lappet, 47, 47a Emperor, 48 Kentish Glory

Plate 9   49 Pebble Hook-tip, 50 Chinese Character, 51 Peach Blossom, 52 Buff Arches, 53 Common Lutestring, 54 Frosted Green, 55 Orange Underwing, 56 March Moth, 57 Grass Emerald

Plate 10 58 Large Emerald, 59 Common Emerald, 60 Little Emerald, 61 Garden Carpet, 62, 62a, 62b Yellow Shell, 63 Dark Spinach

Plate 11 64 Common Marbled Carpet, 65 Grey Pine Carpet, 66 Winter Moth, 67 Twin-spot Carpet, 68 Common Pug, 69 Green Pug, 70, 70a Magpie Moth

Plate 12 71 V-moth, 72 Brimstone Moth, 73 Lilac Beauty, 74 Early Thorn, 75 Swallow-tailed Moth, 76 Feathered Thorn, 77 Pale Brindled Beauty

Plate 13 78, 78a Brindled Beauty, 79, 79a Peppered Moth, 80, 80a Mottled Umber, 81 Waved Umber

Plate 14 82 Willow Beauty, 83 Mottled Beauty, 84 Pale Oak Beauty, 85 Common Heath, 86 Bordered White, 87 Early Moth, 88 Grass Wave

Plate 15 89 Death's Head Hawk-moth, 90 Privet Hawk-moth, 91 Lime Hawk-moth, 92 Eyed Hawk-moth, 93, 93a Poplar Hawk-moth

Plate 16 94, 94a Humming-bird Hawk-moth, 95, 95a Elephant Hawk-moth, 95b Elephant Hawk-moth in defensive attitude, 96 Buff-tip

Plate 17 97 Puss Moth, 97a Puss Moth in defensive attitude, 98 Sallow Kitten, 99 Poplar Kitten, 100 Lobster Moth, 101 Iron Prominent

Plate 18 102 Lesser Swallow Prominent, 103 Swallow Prominent, 104 Coxcomb Prominent, 105 Small Chocolate-tip, 106 Figure of Eight, 107 Vapourer, 108 Dark Tussock

Plate 19 109 Pale Tussock, 110 Brown-tail Moth, 111 Yellow-tail Moth, 112 White Satin, 113 Black Arches, 114 Muslin Footman, 115 Dingy Footman

Plate 20 116 Common Footman, 117 Garden Tiger, 118 Cream-spot Tiger, 119 Clouded Buff, 120 White Ermine, 121 Buff Ermine, 122 Muslin Moth

Plate 21 123 Ruby Tiger, 124 Scarlet Tiger, 125 Cinnabar, 126 Short-cloaked Moth, 127 Garden Dart, 128 Turnip Moth, 129 Shuttle-shaped Dart

Plate 22 130 Flame Moth, 131, 131a Large Yellow Underwing, 132 Lesser Yellow Underwing, 133 Broad-bordered Yellow Underwing, 134 True Lover's Knot, 135 Ingrailed Clay

Plate 23 136 Purple Clay, 137 Setaceous Hebrew Character, 138 Square-spot Rustic, 139 Heath Rustic, 140 Gothic, 141 Red Chestnut, 142 Beautiful Yellow Underwing

Plate 24 143, 143a Nutmeg, 144, 144a Cabbage Moth, 145, 145a Dot Moth, 146, 146a, 146b Bright-line Brown-eye

Plate 25 147, 147a Broom, 148 Antler Moth, 149 Pine Beauty, 150 Blossom Underwing, 151 Common Quaker, 152 Clouded Drab

Plate 26 153 Hebrew Character, 154 Clay, 155 Common Wainscot, 156 Mullein Moth, 157 Sword Grass, 158 Merveille du Jour, 159 Satellite

Plate 27   160 Chestnut, 161 Poplar Grey, 162 Sycamore, 163 Miller, 164 Alder Moth, 164a Alder Moth (early stage), 165 Grey Dagger

Plate 28   166 Marbled Beauty, 167 Copper Underwing, 168 Old Lady, 169, 169a Angle Shades, 170 Dun Bar, 171 Dark Arches, 172 Common Rustic

Plate 29   173 Rosy Rustic, 174 Green Silver Lines, 175 Burnished Brass, 176 Golden Plusia, 177, 177a Silver Y Moth, 178 Red Underwing

Plate 30   179 Mother Shipton, 180 Herald, 181 Snout, 182 Common Fan-foot, 183 Common Swift, 184 Leopard Moth, 185 Goat Moth

Plate 31   186 Forester, 187 Six-spot Burnet, 188 Five-spot Burnet, 189 'Bagworm', 190 Festoon, 191 Currant Clearwing, 192 Common Clothes Moth, 193 Case-bearing Clothes Moth, 194 Small Ermine

Plate 32   195 Diamond-back Moth, 196 Brown House Moth, 197 Pea Moth, 198 Codling Moth, 199 Carnation Tortrix, 200 Green Oak Tortrix, 201 Garden Pebble, 202 Small Magpie Moth, 203 Warehouse Moth

Introduction

Life Cycle
Most people have encountered the hairy 'Woolly Bear' caterpillar wandering across a garden path or the Cabbage White caterpillar feeding in the vegetable plot, and are aware that they are destined to become moths or butterflies. The question that few of us ask is why a moth or butterfly should start its life as a caterpillar. The answer lies in the fact that all insects are encased in an external skeleton made of a tough but flexible material called chitin. This

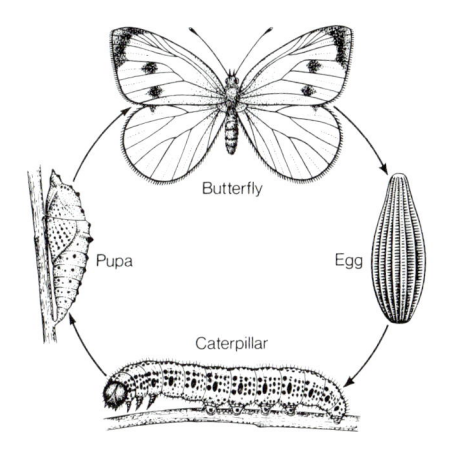

Life cycle of the Large White butterfly

protective outer covering has many advantages, but its serious disadvantage is that it will not stretch sufficiently to allow the animal inside to grow. This problem is solved by shedding the old skin at regular intervals, a process known as moulting or ecdysis. The caterpillar represents the growth stage in the life of a moth or butterfly and is basically a highly efficient, tubular-shaped mobile feeding machine.

The simplicity of the caterpillar's structure makes moulting a relatively easy process, and this takes place a number of times during its development. The number of moults varies between different species, but it is most commonly five or six. The stages between each moult are referred to as instars, the stage emerging from the egg being the first instar. During the caterpillar's growth it forms rudiments of the adult structures within its body, but these are not visible externally. When it is full grown it sheds its skin for the final time and enters the pupa or chrysalis stage.

The pupa stage is the one in which the body cells are reorganized to form the adult moth or butterfly.

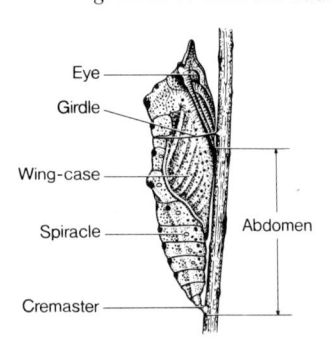

Pupa of the Small White butterfly

Cocoon of the Emperor Moth on heather

During this time the insect is extremely vulnerable as its tissues are easily damaged, and it is immobile and unable to escape predators. Consequently the pupa is enclosed in a strong protective shell and has a 'mummified' appearance, reflected in the scientific name which is derived from the Latin word 'pupa' meaning a doll or puppet. The pupa normally has a group of hooks at the tip of the abdomen forming a structure called the cremaster by which the body is attached to a silken pad spun by the caterpillar. Additionally some butterflies and many moths construct a further protective layer around the pupa called the cocoon. In many cases this is made entirely of silk, but some species incorporate other materials such as leaves, bark or earth. When its development is completed the adult butterfly or moth emerges from the pupa case. The winged adults represent the reproductive phase in the life cycle during which no further growth takes place. Those butterflies and moths which feed do so through highly modified sucking mouth parts capable only of taking up nectar or similar fluids. Many moths do not have functional mouth parts and so are unable to feed at all.

The wings of butterflies and moths perform important functions. The varied and often beautiful patterns formed by the coloured wing-scales are often important in recognition and attraction of mates and, in addition, some of these scales form special scent organs which function to the same purpose. The power of flight is not only important in the seeking of mates but is also essential to the dispersal and distribution of the species. Even small moths can be capable of long flights, and some of our commonest butterflies are regular migrants from southern Europe and North Africa. It should also be noted that some dispersal may take place in the egg stage, as these are sometimes accidently carried by birds or mammals, while small caterpillars are sometimes carried long distances by air currents.

Female butterflies and moths can usually be recognized by their larger and heavier abdomens due to the fact that they may be carrying numbers of eggs. Some female moths are little more than egg-laying machines which have even sacrificed the mobility afforded by wings and remain within or near the cocoon to be sought and fertilized by the winged males.

The life cycle recommences with the process of egg laying. Each egg contains an embryo surrounded by a nutrient fluid which is important for its early development. The outer casing is formed of tough transparent material called chorionin which is often intricately patterned. At the top of the egg is a minute opening called the micropyle (meaning 'a little door') through which the sperm enters during the process of fertilization within the body of the female. After fertilization the micropyle is important in supplying air to the embryo. When the young caterpillar is fully developed and ready to hatch, it bites its way through the egg shell and escapes. Many caterpillars consume the entire shell which forms an important first meal.

Structure of the Caterpillar

Having looked briefly at the life histories of butterflies and moths we can now examine the caterpillar stage in more detail and find out more about the ways in which it is adapted to carry out its basic functions of feeding and growth.

An insect's body is built up of a number of compartments called segments. In certain stages of development, this segmentation may be difficult to detect but, in the caterpillar stage, it is quite obvious except in the case of the head. The caterpillar consists of a head and thirteen body segments. The head consists of six segments intimately fused and contained within a rounded, horny head capsule. The first three body segments, each carrying a pair of jointed legs, make up the thorax while the remaining ten segments form the abdomen.

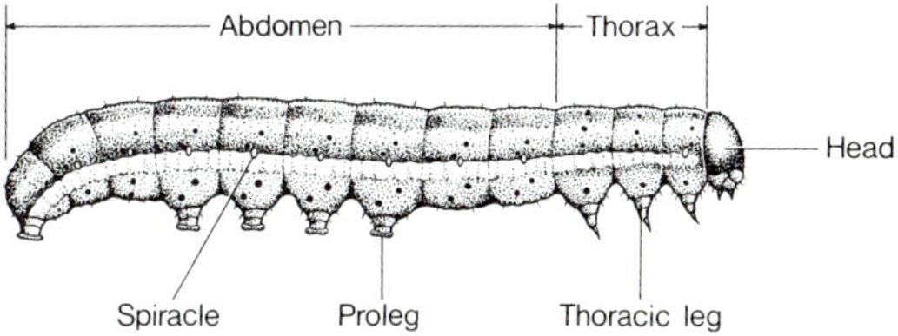

Caterpillar of a Noctuid moth

The mouth parts of the head consist of a pair of robust biting jaws (the mandibles) surrounded by the upper lip (the labrum) and the lower lip (the labium). The labrum is a simple plate carrying sensory hairs, but the labium is more complex as it carries the labial palps and the spinnerets. On either side of the labium are structures called the maxillae. The maxillae and labial palps are sensory organs carrying a number of sensitive hairs. The spinneret is a special outlet of the silk glands which are modifications of the salivary glands. Silk is extruded through the spinneret as a liquid which rapidly hardens to form the fine threads with which we are all familiar. On each side of the mandibles is a segmented antenna which is also sensory in function. The eyes of the caterpillar consist of a group of small lens-shaped ocelli on each side of the head. These ocelli, which usually number six on each side, are probably only capable of distinguishing between light and darkness, and in some caterpillars such as those of the clothes moth they may be absent altogether. Thus the sensory organs of the head such as the antennae and palps assume a most important role.

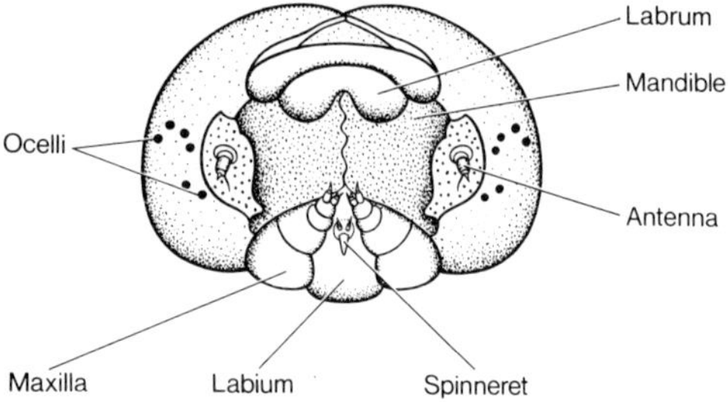

Front view of a caterpillar head showing mouth-parts

The body segments are less complicated and have a fairly uniform structure. Each segment bears a number of hairs or bristles. In some caterpillars these may be small and insignificant while in others they may be more numerous and very prominent. The variety of these hairs and their distribution on the body often reflects the caterpillar's mode of life, and their study is very useful in classification.

On each segment, with the exception of the second and third thoracic and the last abdominal, there is a pair of oval or rounded spiracles. The spiracles represent 'breathing holes' and are the outward openings of a network of tubes called the tracheae. Air diffuses through the spiracles and along the tracheae to all parts of the caterpillar's body. Many caterpillars are able to close the spiracles for a limited time under adverse conditions. In this way they are able to withstand immersion in water, a frequent hazard to caterpillars feeding on trees by rivers and ponds.

Caterpillars have two types of legs: the true, jointed thoracic legs, present on the first three segments, and the unjointed, sucker-like false legs or prolegs usually situated on the third to sixth and on

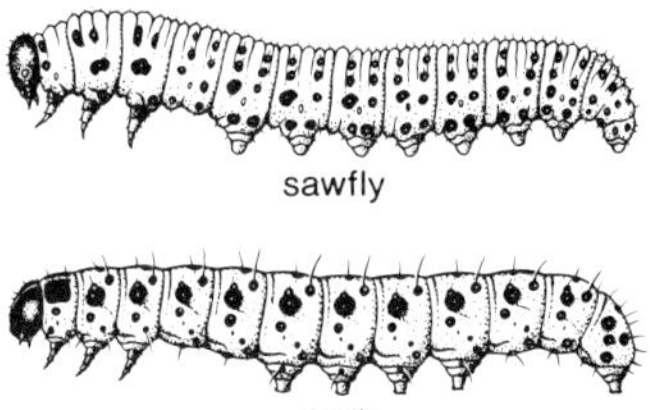

Sawfly and moth caterpillars for comparison

the final abdominal segments. In some caterpillars the number of prolegs is reduced but in no case do they possess more than five pairs. The larvae of sawflies, which look very much like moth caterpillars, usually have more than five pairs of prolegs and always possess one pair on the second abdominal segment, a characteristic by which they may be clearly distinguished.

The prolegs are armed with a number of small hooks which enable the caterpillar to grip securely the surface to which it is clinging. When crawling, the caterpillar makes most use of its prolegs, the segmented thoracic legs being used most commonly in holding food while it is being eaten. By a complicated system of muscular contractions, the prolegs can be extended or retracted in sequence so that the caterpillar is moved along. In the case of looper caterpillars which have a reduced number of prolegs, the method of walking is modified with a more prominent use of the thoracic legs.

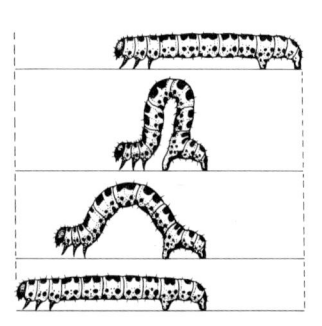

How a looper caterpillar walks

Defence and Camouflage

Caterpillars come in many shapes, colours and sizes and are modified in many ways to suit them to particular modes of life. To return to the Woolly Bear caterpillar, we have just one example of the protective measures adopted. The dense hairs of this and many other caterpillars act as a deterrent to would-be predators and may also protect them to some degree against attack by parasitic wasps and flies. The hairs of some caterpillars are barbed and some are even armed with poisons to add to their defensive role. Amongst British caterpillars with poisonous hairs, the most notorious are those of the Brown-tail Moth which can cause serious discomfort and painful rashes if handled. It is a wise precaution to avoid handling all hairy caterpillars, as many are likely to cause irritation to sensitive skins and can be harmful if the hairs come into contact with the eyes.

Another protective device adopted by certain caterpillars is a gland or group of glands called osmeteria which give off a deterrent smell. Although the odour may not always seem unpleasant to humans it appears successful in warding off predators. The caterpillar of the Puss Moth is capable of ejecting an offensive spray of formic acid at its enemies. In addition to this, the Puss Moth caterpillar can adopt a very impressive threatening attitude when attacked. It rears up with its head drawn back into the thorax and the tails raised and curved forward over the back extruding bright red, whip-like filaments. Another caterpillar which can display an impressive threatening posture is that of the Elephant Hawk-moth. In this case the head is again withdrawn into the thorax which becomes expanded, making the false eye-spots on each side very prominent and giving the insect a snake-like appearance. The Lobster Moth has probably the most unusual caterpillar of all the British moths. Its extremely long legs and tail projections give a spider-like impression when it rears up its head and throws the tail forward in its defence attitude.

Some caterpillars are distasteful to birds and other predators and advertise the fact by the use of warning coloration. The Cinnabar Moth has caterpillars which are conspicuously banded in orange and black and are easily seen when feeding on ragwort plants. The black and yellow caterpillars of burnet moths exude droplets of liquid if disturbed and these droplets contain cyanide compounds which make them unacceptable to most predators.

Apart from these methods of defence, most caterpillars gain protection by being as inconspicuous as possible. Some achieve this by feeding inside plant tissues or underground amongst roots, and some, such as the 'bagworms' (Psychidae), construct portable protective cases which they drag along as they feed. Others are patterned or shaped in such a way that they blend with their background and many are so successful in this that they are seldom noticed until they move. Many looper caterpillars of the family Geometridae show a remarkable resemblance to twigs when at rest. Apart from their coloration, some

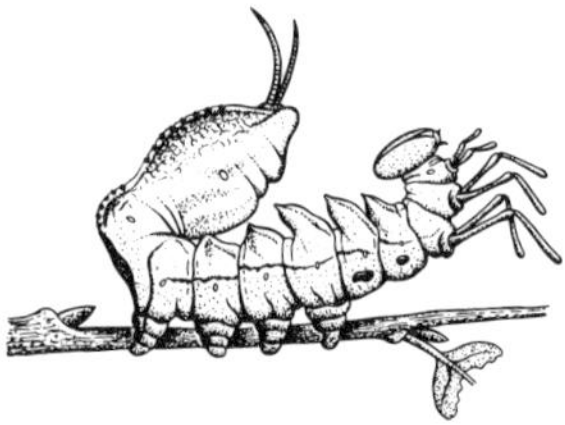

Lobster moth caterpillar in defence posture

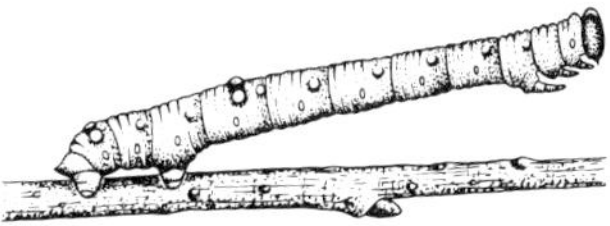

Looper caterpillar at rest on a twig

bear warts and projections which resemble joints, buds or leaf scars.

Caterpillars which feed on pine foliage often have a characteristic green and white striped colour pattern which blends with the needles. Similarly, heather feeders have a broken pattern of green or brown and white which blends remarkably with the small green leaflets and brown stems.

Some caterpillars change their coloration as they grow and become more conspicuous. One interesting example of a drastic change of this nature is that of the Alder Moth caterpillar. Up to the last moult, this caterpillar has a black and white patterning which strongly resembles a bird dropping. In the final instar, however, it is black with conspicuous yellow bands and carries large, club-shaped hairs. It is probable that this warning coloration would not be successful in earlier instars as the caterpillar would be too small to make a significant visual impact. The caterpillars of many hawk-moths are green while they are feeding on the foliage, but when they wander away from the food-plant prior to pupating in the earth their colour darkens and in some cases becomes brown or purplish.

Silk is a very important commodity to the caterpillar. It uses silk to form a supporting platform while moulting and pupating and often spins a shelter of leaves or debris as a protection from predators.

Another use of silk provides an escape mechanism for many caterpillars in that it allows them to drop from a branch when disturbed and remain suspended in mid-air until the danger is past, when they can climb back to their resting place. When suspended on a silk thread some small caterpillars spin rapidly so that they become almost invisible.

Predators, Parasites and Diseases

The most obvious enemies of caterpillars are probably birds and mammals, but smaller predators such as other insects and spiders are a constant danger. Some caterpillars are themselves cannibalistic and will also attack other species. Perhaps the most serious menace from other insects comes from parasitic wasps and flies. Although parasitic wasps are often referred to as 'ichneumons', the family Ichneumonidae represents only one of a large number of parasitic groups. Parasitic flies belong almost exclusively to the family Tachinidae.

These parasites may attack the egg, caterpillar or pupa stage and may feed internally or externally. Some parasites have a sharp pointed egg-laying tube (ovipositor) which can pierce the skin of a caterpillar so that eggs can be deposited within the body tissues. Others lay their eggs on the caterpillar's skin or on nearby leaves, and the young grubs bore into their

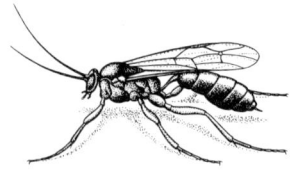

A parasitic wasp

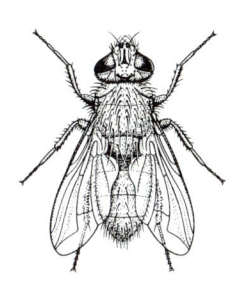

A parasitic fly

hosts or feed externally according to their type. A third method by which caterpillars may be attacked is by swallowing parasite eggs which have been laid on the food-plant.

Caterpillars attacked by external parasites seldom survive to become pupae, sometimes because they are paralysed by the parasite when egg laying but mostly because of the rapid growth of the parasitic grub. Internal parasites often develop more slowly and do less damage to the caterpillar's body in the short term. Some parasitized caterpillars appear to pupate quite normally but are subsequently destroyed by the parasitic grubs which mature within the pupal shell or cocoon. Many internal parasites also become full grown before the caterpillar pupates and these bore their way out of the body before pupating themselves.

Parasites are sometimes valuable in controlling pest species when they reach very high numbers. One of the commonest examples is in the case of the Large White butterfly caterpillars which, when they reach very high numbers, are heavily attacked by parasitic wasps of the genus *Apanteles*. When this happens, shrivelled bodies of the caterpillar hosts may often be

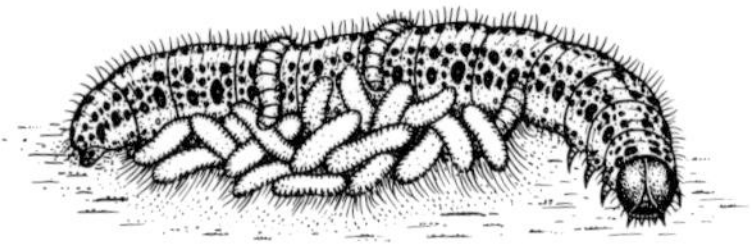

Caterpillar of the Large White butterfly
with cocoons of a parasitic wasp

encountered covered with the characteristic sulphur-yellow cocoons of the parasite.

Caterpillars suffer from a number of diseases, many of which may prove fatal. Both bacteria and viruses are responsible and the symptoms of sickness, usually diarrhoea followed by lethargy and necrosis, are very similar. Some diseases may be passed from one generation to another by a diseased adult producing infected eggs. Parasites are frequently responsible for transmitting diseases while bacterial spores may also be carried long distances in air currents. Fungi are also responsible for some caterpillar deaths, particularly during hibernation. One such fungus attacks caterpillars living underground, invading all the body tissues and then sending out a long fruiting body which looks like a rat's tail. The dried remains of these insects are sometimes called 'vegetable caterpillars' and are regarded as a valuable tonic food in some parts of the world.

Man and Conservation

Perhaps the caterpillar's most deadly enemy is man himself. Our countryside is constantly being changed by man's activities; trees are felled, hedges are grubbed out and vast areas which were once natural downland have gone under the plough to meet our ever-increasing food requirements. Insecticides have caused serious damage to many harmless species of insects, although fortunately we have become aware of the dangers of indiscriminate spraying and a more responsible attitude is now being adopted. Herbicides too have had their effect in destroying the food-plants of many hedgerow species. 'The 'bug-hunter' has come in for heavy criticism for collecting rare insects in numbers likely to affect their chances of survival, but the effects of his depredations have been grossly exaggerated. Whilst we would all hope that collectors will adopt a sensible attitude towards rare or endangered species, few naturalists are likely to deliberately damage the objects of their interest. It is by careful study of the living things which surround us that a useful basis for future conservation may be founded. More and more lepidopterists are becoming reluctant to kill the organisms which they study and prefer to make field observations or to rear eggs in captivity, returning the progeny to the wild. Insect photography is very popular today and is an excellent means of recording one's observations. A good colour photograph can be much more satisfying than a row of dead specimens in a cabinet.

Pests and Their Control

Some caterpillars have taken advantage of man's activities in such a successful way that they are now regarded as serious pests. The most obvious are those which attack field crops and, although their effects are seldom as serious in Britain as they are in the tropics, the damage they cause may amount to millions of pounds each year. Cutworm caterpillars of the family Noctuidae are responsible for most of the field damage in Britain. They are called cutworms because they feed at ground level and often bite through the base of the stem so that the plant falls over. Young plants are seriously affected in this way and even young trees may be killed in numbers. Cutworms will also feed in roots and tubers underground, hollowing them out or holing them so that they are unmarketable. The caterpillars of the Common Swift Moth also feed on roots and sometimes behave as cutworms in gardens, although they feed most commonly on weeds and other wild plants. Their preference for wild plants is the key to their successful control as they seldom thrive in well-cultivated land which is kept free of perennial weeds.

Another group of pests which affect both field and garden crops are the white butterflies. The caterpillars of the Large White butterfly sometimes occur in such numbers that whole fields of cabbages may be stripped so that just the leaf-ribs remain. The caterpillars of the Small White often feed within the hearts of the cabbages and are not noticed until the crop is ruined.

Fruit trees are also the victims of a number of pests, one of the commonest being the Winter Moth which feeds on opening leaf buds in the spring. The Codling Moth caterpillar feeds inside apples and is the common 'maggot' that we all encounter from time to time. Soft fruits are also subject to attack, and the foliage of currant and gooseberry bushes is sometimes seriously damaged by the beautiful black and white caterpillars of the Magpie Moth.

The moths which have adapted themselves most closely to man's environment are the clothes moths and house moths. These are also found in the nests of birds and mammals, and nests in lofts are frequently sources of domestic infestation. It seems however that the advent of man-made fibres has had a serious effect on clothes moths which are much less common in houses today than they used to be.

Some pests are less frequent but, when they do occur, their effect may be quite spectacular. The caterpillars of the Antler Moth feed on grasses, usually on heath or moorland, and they occasionally build up rapidly into great numbers which behave as 'armyworms' devastating large tracts of moorland grazing. Fortunately, these incidents are fairly uncommon, but reports of plagues of caterpillars soon appear in the newspapers when they happen.

Another caterpillar to capture the headlines, especially in the local press of the south east of England, is that of the Brown-tail Moth. The caterpillars may be very abundant where they occur and the foliage of trees and hedges may be stripped. The caterpillars are gregarious and live within tent-like nests which are very conspicuous. It is the poisonous hairs of these caterpillars which make them most newsworthy, especially when children handle them and become covered with painful rashes. An interesting sidelight on this phenomenon may be seen in the writing of William Curtis in 1795. He recounts how villagers were paid by the bushel for collecting nests of this caterpillar. He gives exhaustive accounts of the moth and its effects but makes no mention of the rashes caused by the hairs—a fair indication that our forefathers were somewhat thicker skinned than we are today.

Man has attempted to control insect pests by many means. In the past this was chiefly by the use of chemical insecticides. More recent research not only suggests that such methods may be harmful to man but also that the long-term effects may not be beneficial to agriculture. Careful study of the life cycles of these pests suggests that changes in agricultural practice or biological controls, such as the use of parasites, may give better long-term control. The gardener may not be able to use these methods, but by thorough cultivation and the regular examination of crops and hand picking of caterpillars it may be possible to keep these pests at an acceptably low level without the use of chemicals. Where the use of an insecticide becomes imperative, an approved proprietary brand should be used, taking due care that beneficial insects such as bees and ladybirds are not affected.

Methods of Collecting

While some caterpillars make their presence all too well known, many others are very successful in hiding themselves away and must be looked for very carefully. To start with, a good many caterpillars may be found by searching the foliage of trees, shrubs and low-growing plants, keeping a close look-out for signs of feeding such as holed or spun leaves. Many caterpillars hide away by day and these should be searched for using a torch or lantern at night when they may be found feeding, especially in spring or early summer.

A most productive method of collecting caterpillars is by the use of a beating tray. This may be a specially constructed piece of apparatus or simply an upturned umbrella or similar receptacle. The branches of trees and shrubs are smartly rapped with a stick while the tray is held underneath to catch any caterpillars which may be dislodged. The term 'beating' is unfortunate as it suggests battering the branches and foliage which not only damages the trees but often kills the caterpillars or simply causes them to cling more tightly to their food-plant. Trees that are particularly productive are oak, sallow, hawthorn, blackthorn, and birch, but others are well

Using a beating tray

worth trying. The position of the trees chosen may also be important as those growing on the margins of woodlands are usually better than those in the centre, and isolated trees are sometimes even more productive, particularly if they are not too exposed. Even small shrubby plants may be beaten with success, although caterpillars are usually best collected from low-growing plants by the use of a sweep net. The sweep net is particularly useful for collecting caterpillars of butterflies such as the Meadow Brown, the Gatekeeper and the Speckled Wood when they climb up the grass stems in the evenings. One problem about this method is that it is often difficult to decide which plant the caterpillars were feeding on and this may raise problems if they are to be reared. When caterpillars are collected by any of these methods, it is important to keep them associated with their food-plants. It is advisable to collect from one plant at a time, always bearing in mind that an

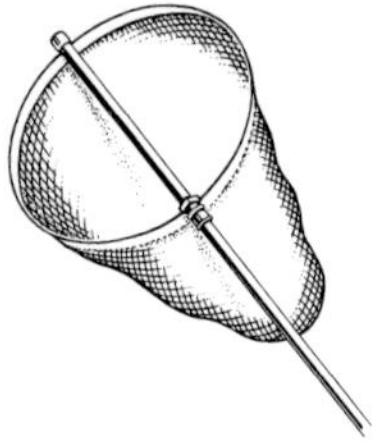

A sweep net

adequate source of food will be needed for the caterpillars once they arrive home. Although a number of different caterpillars may be kept in one container it should be remembered that some are predators; large and small species should be kept apart and a careful look-out should be maintained for the caterpillars of the Dun Bar which should always be isolated immediately.

Plastic sandwich boxes make quite good collecting boxes providing that they are lined with newspaper or other absorbent material to prevent condensation forming, and that the caterpillars are not over-crowded. Glass or plastic tubes may be used for small caterpillars and are useful for isolating suspected cannibals. These tubes should also be lined with absorbent paper to prevent condensation. A small quantity of food-plant should be enclosed in each container which should be fully labelled with locality and date of capture. It is also helpful to carry a notebook, and the more information that can be recorded the greater the ultimate reward and satisfaction.

As the collector gains more experience and has found most of the common species, specialized methods for collecting the rarities or difficult species such as wood-boring clearwing moths may be tried. The key to success in this field is knowledge and experience which may be gained by reading as much as possible and by contact with other entomologists through local or national clubs and societies.

Identification

There are about 2,500 different species of butterflies and moths occurring in the British Isles. The 203 species selected for inclusion in this book have been chosen either because they are widespread and likely to be encountered commonly or because they are of particular interest either for their unusual appearance or specialized biology.

The first clues to the identity of a caterpillar will be its size, shape and colour. A general guide to the shapes and forms of the different families may be found in the silhouettes on the endpapers. This should be followed by checking the plates of the appropriate section until a close match is obtained. At this stage, the text may be consulted to check food-plants, habits and times of appearance. If all of these agree, you may feel fairly confident that you have an accurate identification, but do not expect success every time; remember the other 2,300 species which are not dealt with here. It is always satisfying to confirm one's initial identification by rearing the adult moth or butterfly. The biology of many British moths is still not completely known so any notes, sketches or photographs that you can make of the insect during its development may contribute to our knowledge of the fauna.

Preservation

Caterpillars may be preserved in alcohol or 'blown' and mounted on insect pins in a cabinet. The first method is more useful from the specialist's point of view, as fine structures are better preserved although colour is almost completely lost. Other snags to this method are the cost and restricted availability of alcohol and the difficult storage and display problems. The method of 'blowing' caterpillars appears to be a dying art and one which requires much practice and patience. After killing the caterpillar a cut is made at the tail end and the body contents squeezed out using a small roller. The empty skin is then inflated with air, using a miniature bellows arrangement, and dried in a small oven. Full details of this method will be found in specialist books, but suffice it to say that this is a rather messy and laborious operation which will appeal only to the most dedicated collector. Today's answer to this problem is to use a camera and to record the true beauty of the living insect on film.

Rearing

Many caterpillars are quite easy to rear, and the effort involved is amply rewarded by the sight of a freshly emerged butterfly or moth. Once again the key to success is knowledge gained by experience so it is better to start off with one of the more common species such as the Small Tortoiseshell butterfly or the Angle Shades moth before graduating to those species requiring special treatment. Caterpillars are usually most abundant in the spring and the autumn, and it is a wise move for the beginner to start with spring caterpillars as these usually produce moths in summer. Those found in autumn do not usually produce adults until the following spring, and there is a higher rate of mortality while they are overwintering.

A wide assortment of containers may be used for rearing caterpillars, ranging from simple jam jars to purpose-built cages of wood, glass and netting.

Plastic sandwich boxes are extremely valuable for rearing small caterpillars and can be used for larger ones providing they are not overcrowded. Condensation is often a problem in such containers, but this may be overcome by lining them with sheets of newspaper or other absorbent material and, where necessary, by boring air holes in the lid. Caterpillars require very little air and will survive in a closed box for some time. It is better not to bore air holes in the lids of boxes housing very small caterpillars as they are likely to escape or become dehydrated. The secret of success when rearing caterpillars in such containers is to avoid overcrowding and to maintain a high standard of hygiene. Food should be changed regularly, and any frass (droppings) or other debris removed. The absorbent paper liner should also be replaced when soiled, and any traces of condensation should be removed. When moving caterpillars to fresh food it is best to place this in the container and allow them to wander on to it. If necessary they may be moved using a soft brush but should never be handled as they are easily injured. Caterpillars should never be disturbed when they are moulting.

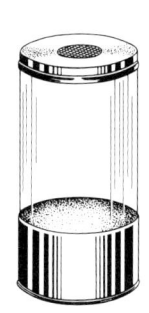

A cylinder cage

The most popular containers used by the beginner are probably jam jars and shoe boxes, both of which have certain drawbacks. The problem of the jam jar is one of cleaning as it is very difficult to remove frass and wilted food without disturbing the inmates. Shoe boxes have the disadvantage that they are difficult to make caterpillar-tight and they allow the food-plant to dry up too quickly. With care both types of container may be used with success, but the following simple cages are likely to produce better results.

The cylinder cage is a popular design which has been sold by dealers for many years and is still obtainable today. Similar cages may be constructed at home using tins of various sizes and transparent tubes made of perspex or plastic sheeting of various types. Another similar type of cage is also sold by dealers but may be constructed by the DIY enthusiast with little trouble. This consists basically of a wooden box covered at the back with nylon gauze and at the front

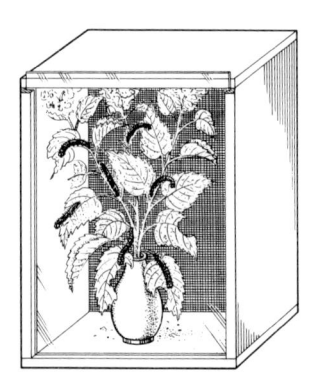

A box cage

by a sheet of glass which can be slid out when the cage is to be cleaned.

Where the size of the cage permits, the stems of food-plant should be put into a small jar of water plugged with cotton wool at the neck to prevent the caterpillars from drowning. It is better still to use potted plants as these will last much longer. It should be remembered that there are restrictions governing the digging up of wild plants, and care should be taken to comply with regulations. When it is not practical to use growing plants or even to supply them with a container of water, the foliage must be replenished as soon as it shows signs of wilting or becoming fouled with frass. In some cases, caterpillars seem to prefer wilted to fresh foliage, but such instances must be treated on their own merits when they arise.

One final method of rearing which can be applied to many species which feed on trees and shrubs is that of 'sleeving'. The 'sleeve' comprises a wide tube of muslin or nylon netting which is slipped over a branch and tied at both ends after the caterpillars have been introduced. Often the caterpillars may remain in this sleeve until they are full grown, but if the foliage is consumed, or overcrowding occurs, they must be moved to another branch. This method

A sleeved branch

has the advantage that the caterpillars need little attention, but it does require access to trees that will not be disturbed by others and does not permit the detailed observations that can be made with an indoor cage. This is the technique so successfully adopted by the butterfly farmers when rearing caterpillars for sale.

When the caterpillars become full grown, the next problem is to provide them with a suitable place in which to pupate. Pupation habits vary greatly between different species and, if possible, these should be ascertained before the caterpillars are full fed. Many will pupate either among the food-plants or between the layers of lining paper provided, but those which usually pupate below ground should have their cages lined with a suitable depth of peat or sphagnum moss into which the caterpillars can burrow. The first sign that a caterpillar is about to pupate is when it becomes restless and wanders about the cage in search of a suitable pupation site. It then becomes sluggish and foreshortened and sometimes changes colour, usually becoming darker and less clearly marked. Those specimens which spin cocoons will of course be quite obvious, but care should be taken that they do not make their cocoons where they will be damaged when the cage is opened. This is a critical stage in the life cycle and the insects should not be disturbed until the pupa is fully formed and has had time to harden.

Those pupae which will emerge within a few months may usually be left in the rearing cage providing that there is room for the butterfly or moth to emerge without damaging itself. Others remain in the pupa for some months, particularly those which overwinter, and these should usually be removed and stored in an air-tight tin or plastic box in an unheated shed or in a sheltered part of the garden. This method ensures that the pupae do not dry out but at the same time are not damp enough to allow the formation of moulds. Some species seem to be highly resistant to mould attacks but others are frequently killed if kept too damp. When the time for the moths to emerge draws near, the pupae should be transferred to a suitable emergence cage. The pupae should be placed on peat or sphagnum moss to prevent them from rolling; some collectors place them between the corrugations of a sheet of corrugated cardboard, which serves this purpose very well. At this stage they may be lightly sprayed with a fine mist of water each day. Those pupae which are suspended from silken girdles or spun up in cocoons should not be moved or disturbed unless absolutely necessary as this often results in the emergence of crippled adults.

Failures will of course occur, but one should learn from these mistakes and modify techniques accordingly. If a parasite emerges it should not be killed but observed and recorded as an integral part of the life history. There are still many gaps in our knowledge of the parasites and these would repay a special study in themselves.

All cages and containers should be thoroughly cleaned and sterilized before re-use to prevent the transmission of diseases. Plastic boxes may be sterilized in a solution of a proprietary disinfectant, but this should not be too strong or it may attack the plastic. Wooden cages may be sterilized in a solution of washing soda.

A final point to be considered has already been mentioned in passing, but it is of paramount importance. We must all be aware of the pressures that are being brought to bear on our wildlife and should take every care to ensure that we do not make matters worse. It is highly questionable whether collectors have ever seriously affected insect populations in this country, but everyone should take a responsible attitude and cause as little disturbance to wildlife as possible when collecting or observing. Before collecting caterpillars by any of the methods mentioned in this book care should be taken to obtain the permission of landowners, and it should be remembered that this applies to all land owned by the National Trust and such bodies as the Forestry Commission. When caterpillars are reared from egg batches in captivity, the numbers of adults obtained are usually much higher than they would be in the wild, indeed one moth may produce over 500 progeny in this way, whereas in nature it is only expected to produce one pair of mature offspring. Many collectors have taken advantage of this fact to 'reinforce' populations of insects by breeding and releasing them. This admirable idea may however cause problems, particularly if the insects are not released in the same locality from which the parents came. Even the correct locality could be swamped by too heavy reinforcement, and insufficient food-plant would be available to support the resulting caterpillars. It is always a difficult problem to try to redress the balance of nature artificially but, given expert advice, it may be possible for the amateur to make a useful contribution in this way. If such a project is undertaken it should be in consultation with the appropriate County Naturalist's Trust or any other such interested body within the area.

It is impossible to cover the field of observing and studying caterpillars comprehensively in this brief introduction, and those who wish to study this fascinating subject in more detail should consult the books listed for further reading.

Notes

Food-plants The food-plant lists are not completely comprehensive but give the more usual food-plants, generally in rough order of preference although this is often difficult to ascertain. Where caterpillars have been successfully reared in captivity on substitute food-plants, these have been mentioned.

Season This indicates the time of year at which the caterpillars may be found. The caterpillars of a few species may be found at some stage of development in every month of the year.

Magnification Most of the illustrations are approximately life size, although a few of the larger species have been slightly reduced and smaller species such as the Geometridae have been enlarged. In these cases a guide to the magnification is given on the plate.

Nomenclature Scientific nomenclature follows Kloet & Hincks' revised Check List of British Insects, part 2 Lepidoptera 1972.

BUTTERFLIES

1 The Small Skipper Butterfly

Family HESPERIIDAE *Thymelicus sylvestris*

Food-plants Grasses such as cat's-tail, creeping soft-grass and Yorkshire fog
Season August–June
Distribution May be found in most parts of England and Wales as far north as Lancashire and Yorkshire
Notes On hatching, the small caterpillar eats part of its eggshell and then spins a silken cocoon in which it remains until the following April. It then starts feeding on a blade of grass, concealing itself by spinning the edges together with silk to form a tube. The colour and pattern of the full-grown caterpillar make it difficult to see on the food plant. When fully grown it joins several leaves together with silk to form a shelter and turns into a pupa which is also green. The butterfly emerges from the pupa in late June or early July.

2 The Large Skipper Butterfly

Family HESPERIIDAE *Ochlodes venata*

Food-plants Various grasses, including cat's-tail, cock's-foot, Yorkshire fog and couch grass
Season July–May
Distribution Widespread in England and Wales and also occurs in southern Scotland
Notes The newly hatched caterpillar spins the edges of a grass blade together to form a tube in which it lives, crawling out only to feed. In autumn the caterpillar constructs a silken shelter where it spends the winter, becoming active again in spring. When fully grown in May it makes a cocoon in the fold of a grass

blade in which it pupates. The pupa is dull black with a grey-green abdomen and the butterfly emerges after about three weeks.

3 The Dingy Skipper Butterfly

Family HESPERIIDAE *Erynnis tages*

Food-plants Bird's-foot trefoil and horseshoe vetch. In captivity the caterpillar will eat the foilage of garden strawberry
Season June–May
Distribution Widely distributed in Great Britain but more common in chalk and limestone areas. It is very local in Ireland
Notes On hatching, the caterpillar spins a few leaflets of the food plant together and feeds within the shelter formed. When it has eaten the inner surfaces of the leaflets it then constructs a fresh shelter in the same manner. The caterpillar becomes fully grown by about the middle of August when it spins a silken cocoon-like structure among the leaflets. It then hibernates through the winter before pupating in April or early May without any further feeding. The slender pupa is greenish with a brown abdomen, and the butterfly hatches after four or five weeks.

4 The Grizzled Skipper Butterfly

Family HESPERIIDAE *Pyrgus malvae*

Food-plants Wild strawberry, bramble, wild raspberry and cinquefoil. This caterpillar will also eat garden strawberry in captivity
Season June–August
Distribution Widely distributed in England and Wales as far north as Yorkshire
Notes After emerging from the egg, the caterpillar

spins a silk covering over itself, usually while resting along the mid-rib of a leaf. It crawls to about its own length out of its shelter to feed on the surface of the leaf and then goes back into hiding. Later it draws the edges of the leaves together and lives within the folds. When fully grown in autumn the caterpillar pupates in a folded leaf. The pupa is brown with greenish wing-cases, and the butterfly hatches in the following spring.

5 The Swallow-tail Butterfly

Family PAPILIONIDAE *Papilio machaon*

Food-plants Milk parsley, angelica, fennel and wild carrot. In captivity the caterpillar can be reared on the foliage of carrot or parsnip
Season May–August and September–October
Distribution Low-lying fenland in Norfolk, Suffolk and Cambridgeshire
Notes This butterfly is double-brooded in favourable years, the first brood caterpillars hatching in late May or early June and pupating in August and the second brood hatching in August and pupating in October. When the caterpillar hatches it resembles a bird-dropping, being black with a conspicuous white patch in the middle of the body. After the third moult it assumes its characteristic green colour. When fully grown the caterpillar pupates either on the food plant or on an adjacent reed stem, attached to a silken pad by its cremaster and supported by a silken girdle. The pupa is very variable in colour and can be pale green, brownish or even black. Under favourable conditions some of the pupae of the first brood emerge in August to produce the second brood, while the others overwinter and emerge in the following spring together with those of the second brood.

6 The Clouded Yellow Butterfly

Family PIERIDAE *Colias croceus*

Food-plants Clover, lucerne and bird's-foot trefoil
Season June–July and September–October
Distribution This migrant butterfly has been recorded from all parts of the British Isles, but is more frequently seen along the south coast of England
Notes Caterpillars found in June are the offspring of butterflies which have migrated here in the spring. They pupate in July, and the resulting butterflies produce a second brood of caterpillars in September which sometimes pupate in October but do not survive the winter. For the first week of its life the caterpillar feeds on the surface layer of a leaf, but after its first moult it eats right through the leaves, making holes in them. The green pupa is attached by its cremaster to a pad of silk spun on the stem or leaf of a plant and is supported by a silken girdle.

7 The Brimstone Butterfly

Family PIERIDAE *Gonepteryx rhamni*

Food-plants Buckthorn and alder-buckthorn
Season June–July
Distribution Throughout England and Wales south of Cumbria, also occurring in Ireland where it is more common in the south and west
Notes The caterpillar rests along the mid-rib of a leaf which it matches in colour and is very difficult to see. It can often be detected by its habit of eating the tip and upper part of the leaf on which it is resting. When fully grown the caterpillar usually leaves the food-plant and often wanders for some distance before pupating amongst rough herbage. The green pupa has enlarged wing-cases and resembles a curled leaf. It is attached to a pad of silk on a stem or leaf

by the hooks of its cremaster and supported by a fine silken girdle. The butterfly emerges at the end of July or the beginning of August.

8 The Large White Butterfly

Family PIERIDAE *Pieris brassicae*

Food-plants Plants of the cabbage family and also nasturtium and mignonette
Season July and September–October
Distribution Throughout the British Isles
Notes This caterpillar is of economic importance as it sometimes causes severe damage to field crops of cabbages and other related vegetables. It is also a garden pest and is frequently found on nasturtium. The caterpillars of this butterfly are unusual in emitting an unpleasant and characteristic smell. There are two broods in the year, the first hatching in early July and pupating, usually on the food-plant, in late July. The second brood caterpillars hatch in September and pupate in October, choosing sites such as fences, walls or window frames. The pupae of the first brood emerge after about two weeks whilst those of the second brood remain in this stage over winter, emerging in the spring. The colour of the pupa varies considerably depending on its surroundings. The normal colour is ivory or greenish white marked with black, but when attached to the food-plant it is some shade of green. The pupa is attached to a pad of silk spun by the caterpillar and supported by a silken girdle around the middle.

9 The Small White Butterfly

Family PIERIDAE *Pieris rapae*

Food-plants Plants of the cabbage family and also nasturtium and mignonette

Season June–September
Distribution Throughout the British Isles
Notes Like the Large White, this caterpillar is of economic importance and is frequently a pest of cabbage crops. There are two or three broods a year depending on weather conditions. The first brood caterpillars are fully fed and pupate at the end of June to produce butterflies in July. The second brood caterpillars hatch in July and pupate after about three weeks. In warm summers butterflies emerge in August and lay eggs which hatch after a few days to produce third brood caterpillars. These become fully fed and pupate in the autumn, and the pupae overwinter before emerging in spring. In cool summers the pupae of the second brood overwinter. The pupae of the spring brood are often attached to leaves and are then always green. Those of later broods are normally found on walls, fences or the sides of buildings and are variously coloured from cream-white to dull grey to merge with their surroundings. Whatever pupation site is chosen, the pupa is attached to a basal pad of silk by its cremaster and supported by a silken girdle around the middle.

10 The Green-veined White Butterfly

Family PIERIDAE *Pieris napi*

Food-plants Hedge mustard, garlic mustard, charlock, watercress, horse-radish and other related plants
Season June–September
Distribution Throughout the British Isles
Notes This butterfly is double-brooded. The first brood caterpillars pupate in June and July. Some of the resulting butterflies emerge in July and August whilst others overwinter as pupae and emerge in early spring. The summer butterflies produce second brood

caterpillars which pupate in the autumn. These pupae also overwinter and the butterflies hatch in spring. The full-grown caterpillar usually chooses a pupation site amongst the foliage of the food-plant, but sometimes chooses fences or similar structures. The pupa attaches itself to a silken pad by means of its cremaster and is supported by a silken girdle around the middle. Its colouring is very variable ranging from bright green to pale buff, either without markings or strongly marked and speckled with black.

11 The Orange-tip Butterfly

Family PIERIDAE *Anthocharis cardamines*

Food-plants Hedge mustard, garlic mustard, lady's smock, horse-radish, charlock, watercress and other related plants

Season June–July

Distribution Generally distributed throughout England and Wales and in Scotland as far north as Inverness-shire. Widespread and common in Ireland.

Notes On hatching, the young caterpillar eats the empty egg shell and will also eat any unhatched eggs that it finds. It then feeds on seed pods of the food-plant which it closely resembles in pattern and coloration. In the early stages the caterpillars tend to be cannibalistic and so should be isolated when reared in captivity. When fully fed, the caterpillar wanders away to find a suitable site to pupate, usually in a hedgerow or amongst dense vegetation. The colour of the pupa varies from green to pinkish brown. It is attached to a silken pad, spun on a branch or stem, by its cremaster and is supported by a silken girdle. The pupa stage usually lasts from ten to eleven months, although in captivity some may pass a second winter before emerging.

12 The Green Hairstreak Butterfly

Family LYCAENIDAE *Callophrys rubi*

Food-plants Gorse, broom, bird's-foot trefoil, bilberry, buckthorn, bramble, heather and many other plants
Season June–July
Distribution Widespread throughout the British Isles although more common in Ireland
Notes When feeding on the fruits of buckthorn holes are made in the berries through which the contents are extracted. Buds of bramble are attacked in a similar way. After the first moult the caterpillars become cannibalistic, often attacking smaller individuals when they are about to moult. They retain this habit until fully grown and so should be isolated when reared in captivity. When full grown the caterpillar leaves the food-plant and pupates under litter on the ground. The pupa is brown with black markings and is spun over by a few silken threads. This stage lasts for about ten months.

13 The Purple Hairstreak Butterfly

Family LYCAENIDAE *Quercusia quercus*

Food-plants Usually oak, but has been found on sallow and sweet chestnut
Season April–June
Distribution Widespread in England and Wales and parts of west Scotland. Very local in Ireland
Notes The caterpillars hatch as the oak buds are bursting and start to feed on the tender foliage. They spin silk over the stems, bracts and leaf bases to form a shelter in which to feed. Caterpillars may be collected by beating the branches of isolated oak trees in favourable localities. The full-grown caterpillars pupate under litter on the ground or in a crevice

in the bark. The pupa is reddish brown with dark markings and is covered by a few strands of silk. The butterfly emerges after about one month.

14 The Small Copper Butterfly

Family LYCAENIDAE *Lycaena phlaeas*

Food-plants Sorrel and dock
Season May–June, August–September and October–March
Distribution Widespread and often common throughout the British Isles
Notes This butterfly is triple brooded. Caterpillars of the third brood hibernate through the winter on the food-plant and feed again in the spring before pupating. In the hibernating phase the caterpillar changes from bright green to dull olive green, reverting to bright green again in the spring. Some caterpillars have red markings which closely match the reddish discoloured areas often seen on the leaves of dock and sorrel. The pupa is brown and is attached to a silk pad, spun on the food-plant, by the cremaster and supported by a silken girdle. The pupa stage of the first two broods lasts for about one month.

15 The Small Blue Butterfly

Family LYCAENIDAE *Cupido minimus*

Food-plant Kidney vetch
Season June–May
Distribution Widespread but local in England, Wales and Scotland. Widespread and often common in Ireland, particularly in limestone areas of the west
Notes Soon after hatching from the egg which is laid on the flower of the food-plant, the caterpillar bores through the calyx to feed on the developing

seeds. It becomes full grown in July and then goes into hibernation until the following May when it pupates. The hibernating caterpillars match the dead calyces of the food-plant so well that they are almost impossible to find. The pupa is greyish white freckled with brown and is attached head upwards to a grass blade by its cremaster and silken girdle. The butterfly emerges after about two weeks.

16 The Common Blue Butterfly

Family LYCAENIDAE *Polyommatus icarus*

Food-plants Bird's-foot trefoil, rest harrow and clover
Season June–August and September–April
Distribution Generally distributed and often common throughout the British Isles
Notes This butterfly is usually double brooded in the south and single brooded in the north. In favourable seasons there may be a third brood in the south. The newly hatched caterpillar is grey in colour but gradually changes to bright green during its development. Those which hatch in the autumn hibernate when quite small and start feeding again in March or April. The caterpillars change to dull olive green during hibernation, resuming their bright green colour in the spring. When full fed the caterpillars pupate at the base of stems of the food-plant in a slight cocoon. The pupa is green with a brownish head and wing-cases, this stage lasting for a fortnight.

17 The Chalkhill Blue Butterfly

Family LYCAENIDAE *Lysandra coridon*

Food-plants Mainly horse-shoe vetch, but also recorded on bird's-foot trefoil, kidney vetch and bird's-foot
Season April–June

Distribution Fairly common in chalk and limestone areas in southern England, but has been recorded as far north as Cumbria

Notes The caterpillars hatch in spring from eggs laid in the previous autumn. At first they feed on either the upper or under surfaces of the leaves, but later eat all parts of the plant including the stems. The caterpillars feed mostly at night, sheltering at the base of the plant by day. Like many other caterpillars of the family Lycaenidae, those of the Chalkhill Blue possess a 'honey gland' on the abdomen which secretes a sweet fluid attractive to ants. The ants feed on this fluid and sometimes move the caterpillars to plants nearer their nests. In return, the caterpillars are protected by the presence of the ants; an interesting case of symbiosis. When full grown, the caterpillar crawls to the base of the plant where it pupates in a crevice in the earth or amongst the roots. The pupa is ochreous yellow inclining to greenish on the thorax. The butterfly hatches after about a month.

18 The Holly Blue Butterfly

Family LYCAENIDAE *Celastrina argiolus*

Food-plants Holly, dogwood, alder buckthorn, spindle and furze in spring. Ivy and sometimes bramble in autumn

Season June and August–October

Distribution Widespread and common throughout southern England and Wales, but occurring as far north as Dumfriesshire and also in Ireland

Notes This butterfly is double brooded in southern England and Ireland but single brooded in the north. Caterpillars of the first brood which hatch in the spring usually feed on the flower buds and unripe berries of holly. Those of the second brood usually feed on ivy, eating the buds and young berries. The caterpillars feed by night and are very sluggish, not wandering far from the place where they hatched. The full-fed caterpillar attaches itself to a leaf where it pupates. The stout pupa is pale brown with darker markings. Those of the spring caterpillars produce butterflies after about a fortnight, but the autumn pupae overwinter to produce butterflies in the following spring.

19 The White Admiral Butterfly

Family NYMPHALIDAE *Ladoga camilla*

Food-plants Honeysuckle. In captivity will eat the foliage of snowberry
Season August, September–May or June
Distribution Generally confined to woodland localities in southern and eastern England, although it is also recorded from Devon, Shropshire and Worcester
Notes The young caterpillar has a remarkable method of camouflaging itself. It attaches pellets of its own droppings mixed with silk to its body so that it resembles a fragment of debris. In autumn, the small caterpillar constructs a winter shelter by spinning a honeysuckle leaf against a twig to form an enclosed chamber. It leaves its shelter in the following spring when it recommences feeding on the new foliage. When full fed the caterpillar suspends itself head downwards from a stem of the food-plant and pupates in this position. The handsome angular pupa has two ear-like projections from the head. The ground colour is usually green with olive brown markings beautifully decorated with metallic spots and a golden sheen. The butterfly hatches after about a fortnight.

20 The Purple Emperor Butterfly

Family Nymphalidae *Apatura iris*

Food-plants Sallow. In captivity will eat poplar and willow
Season August–June
Distribution Local in larger oak woods in southern England. It has also been recorded from Wales
Notes The newly hatched caterpillar does not possess the pair of horns which are so conspicuous when it is full grown. The horns appear after the first moult, and these are held straight in front when the caterpillar rests along the midrib of a leaf so that it merges with the background. In October it spins a pad of silk at the base of a branch or on a leaf to which it attaches itself while in hibernation. During hibernation the caterpillar becomes brownish but resumes its green colour when it recommences feeding in the spring. When full grown the caterpillar spins a pad of silk to the underside of a leaf which is itself secured to its stem by silken threads. The pupa which is hump-backed and pale green hangs head downwards from the silk pad. The butterfly emerges after about a fortnight.

21 The Red Admiral Butterfly

Family Nymphalidae *Vanessa atalanta*

Food-plants Usually stinging nettle, but has been found on pellitory and hop
Season June–August
Distribution This migrant butterfly is found throughout the British Isles
Notes The caterpillars hatch from eggs usually laid by butterflies which have migrated here in the spring. Each caterpillar constructs a tent-like shelter by spinning together the leaves of its food-plant which it then

proceeds to eat. The basic colouring of the caterpillar is very variable, ranging from velvety black to greenish grey. When full fed it constructs a pad of silk on the undersurface of a leaf surrounded by a canopy of foliage spun together to form a shelter. It then attaches itself to the silken pad and pupates head downwards. The angular pupa is pale brown covered with a powdery grey bloom and decorated with golden metallic patches. The butterfly hatches after two or three weeks.

22 The Painted Lady Butterfly

Family NYMPHALIDAE *Cynthia cardui*

Food-plants Thistles, burdock, viper's bugloss, mallow, stinging nettle, and has been found on runner beans
Season June–July
Distribution This migrant butterfly has been found in all parts of the British Isles
Notes The caterpillar lives and feeds within the shelter of a silken web spun across a leaf. As it feeds and grows, larger shelters are constructed and several discarded dwellings may be found on one plant. The colour of the spines of this caterpillar vary from completely pale yellow to black with a yellow band. The full-fed caterpillar spins a silken pad on a stem and pupates head downwards. The pupa is generally brownish grey with darker brown markings and burnished metallic patches. The butterfly emerges after about a month.

23 The Small Tortoiseshell Butterfly

Family NYMPHALIDAE *Aglais urticae*

Food-plant Stinging nettle
Season May and July–August

Distribution Common and widespread throughout the British Isles

Notes This butterfly is double brooded. When small, the caterpillars are gregarious, living together under a communal web spun over the terminal leaves of the food-plant. They move off in a body to a fresh plant when the young leaves have been consumed. In the final stage the caterpillars disperse, and when full fed often wander considerable distances to find a suitable pupation site. The pupa hangs head downwards from a silken pad spun on any suitable support, often on fences and palings. The slender pupa is grey, often with a pinkish suffusion and has metallic patches on the body. Butterflies of both broods hatch after about twelve days.

24 The Peacock Butterfly

Family NYMPHALIDAE   *Inachis io*

Food-plants Usually stinging nettle, but will also eat hop

Season June–July

Distribution Widely distributed in the British Isles except northern Scotland

Notes At first the young caterpillars are greenish grey and hairy. They are gregarious, living and feeding in a mass under a web spun over the leaves of the food-plant. As they grow they become velvety black and develop spines. When full fed, the caterpillars seek a suitable pupation site, sometimes on the food-plant, but often some distance away. The pupa is suspended head down from a silken pad. Its colour may be pale yellow green, pinkish grey or olive brown, but it is usually speckled with black. There are two pointed horns on the head. The butterfly emerges after about a fortnight.

25 The Comma Butterfly

Family NYMPHALIDAE *Polygonia c-album*

Food-plants Usually stinging nettle, hop or currant, but will also eat gooseberry, elm and sallow
Season April–June and August
Distribution Throughout central and southern England and Wales. It does not occur in Ireland
Notes This butterfly is double brooded. On emerging from the egg, the young caterpillar spins a slight web on the underside of a leaf before commencing to feed. At this stage it is greenish in colour and does not possess spines which develop later. The caterpillars are usually solitary, but when kept in captivity will feed together in small groups. Those of the first brood moult four times, but those of the second brood only have three moults. The full-grown caterpillar is black with a large patch of creamy white on its back so that when at rest it strongly resembles a bird dropping. When fully fed it pupates on the food-plant. The pupa is suspended head downwards from a pad of silk. It is brownish, suffused with pink and with metallic silvery or golden spots on the body. The butterfly emerges after ten to twenty days according to weather conditions.

26 The Small Pearl-bordered Fritillary

Family NYMPHALIDAE *Boloria selene*

Food-plants Usually dog violet, but possibly other violets as well. In captivity will eat garden viola and pansy
Season July–May
Distribution Locally common in woodlands throughout England, Wales and Scotland, but not recorded from Ireland
Notes The young caterpillars are pale olive green

with brown warts and a black head. After feeding for two months, the caterpillar shelters in the curled-up portion of a withered violet leaf, remaining there until the following spring. It then emerges to complete its feeding. When full fed it pupates on the food-plant hanging head downwards from a silken pad spun on a leaf or stem. The pupa is brown with black markings and a number of metallic spots on the sides of the body. The butterfly emerges after about a fortnight.

27 The Pearl-bordered Fritillary

Family NYMPHALIDAE *Boloria euphrosyne*

Food-plants Violets, usually dog violet. In captivity will eat garden viola and pansy
Season June–April
Distribution Locally common in woodlands in England and Wales, but in Scotland is less common than the Small Pearl-bordered Fritillary. In Ireland it occurs only in the Burren, Co. Clare
Notes When first hatched the caterpillar is pale green, gradually becoming darker as it matures. At the end of July the small caterpillar goes into hibernation, usually on the underside of a withered leaf. In the following March it leaves its shelter and recommences feeding. When full fed the caterpillar pupates on the food-plant, suspended from a silk pad spun on a stem or the underside of a leaf. The pupa is grey with a network of brownish markings. The butterfly emerges after about ten days.

28 The High Brown Fritillary

Family NYMPHALIDAE *Argynnis adippe*

Food-plants Dog violet and sweet violet. In captivity will eat garden viola and pansy

Season April–June
Distribution Large woods throughout England and Wales, but less common in the north. Not recorded from Scotland or Ireland
Notes The young caterpillars vary in colour from green to brown, developing spines after the second moult and gradually becoming darker. When full fed, the caterpillar constructs a tent-like shelter by spinning a few leaves together before pupating. The pupa is suspended head down from a pad of silk usually spun on the stem. It is brown with paler speckling and bears two rows of blunt spines along the back which are a bright greenish-gold colour. The butterfly emerges after about a month.

29 The Dark Green Fritillary

Family NYMPHALIDAE *Argynnis aglaja*

Food-plants Dog violet and probably other violets. In captivity will eat garden viola and pansy
Season August–June
Distribution Locally common throughout the British Isles. In Ireland it is mostly confined to the coastal regions
Notes The caterpillars hibernate soon after they have hatched from the egg in August, sheltering at the base of the food-plant or under dead leaves. They do not become active again until the following spring and have been observed to live without feeding for 228 days. When full fed in June, the caterpillar spins leaves together to form a tent-like shelter within which it pupates suspended head downwards from a silken pad attached to a stem or leaf. The pupa is black with pale brown markings and a brown abdomen. The butterfly emerges after about a month.

30 The Silver-washed Fritillary

Family NYMPHALIDAE *Argynnis paphia*

Food-plants Usually dog violet, but possibly other violets as well. It has been recorded from wild raspberry and will feed on garden viola and pansy in captivity

Season August–May

Distribution Locally common in southern England, South Wales and Ireland. In northern England and North Wales it is scarce, and there are doubtful records from Scotland

Notes On emerging from their eggs, which are usually laid on tree-trunks near to the food-plant, the young caterpillars shelter in crevices in the bark and hibernate until the following spring. They commence feeding in late March or early April and become full fed in May. The pupa is suspended head down from a silk pad spun on a twig or other suitable support. It is pale brown with darker markings and metallic points on the body which resemble droplets of moisture shining on a dead leaf. The butterfly emerges after about two and a half weeks.

31 The Speckled Wood Butterfly

Family SATYRIDAE *Pararge aegeria*

Food-plants Various grasses, including annual meadow-grass, couch grass and cock's-foot

Season Throughout the year

Distribution Widespread and common in southern England and Wales, becoming more scarce further north. In Scotland it is mainly confined to the west coast. It is generally distributed and common in Ireland

Notes This butterfly is generally double brooded in the British Isles, but these broods are divided into early and late phases. This is brought about by the fact that caterpillars hatching in late summer can either pupate in autumn or overwinter as caterpillars and pupate in the spring. The former will produce butterflies in April whereas the latter produce butterflies in June. In this way caterpillars can be found in various stages of development throughout the year. When full fed the caterpillars pupate hanging head downwards from silken pads spun on blades of grass. The pupae are usually green, but a brown form also occurs. Butterflies emerge after about a month, except in the case of overwintering pupae where they emerge after about six months.

32 The Wall Butterfly

Family SATYRIDAE *Lasiommata megera*

Food-plants Various grasses, including annual meadow-grass, couch grass and cock's-foot
Season September–April and June–July
Distribution Widespread and often common in England, Wales and Ireland. In Scotland it is confined to the south west
Notes This butterfly is double brooded in the British Isles, and in favourable years there may be a third brood in the autumn. Caterpillars hatching in spring and early summer produce butterflies in July and August. The second brood caterpillars hatch in autumn, and most of these hibernate through the winter, recommencing their feeding in spring and producing butterflies in May and June. However, it appears that some of these caterpillars may continue to feed through the winter and produce butterflies as early as April. The caterpillars feed at night on the grass blades. When full fed they pupate hanging head downwards from a silken pad spun on the grass stems. The pupae vary in colour from bright green to black

with white or yellow spots on the body. The butterflies emerge after about a fortnight.

33 The Marbled White Butterfly

Family SATYRIDAE *Melanargia galathea*

Food-plants Grasses such as sheep's fescue, annual meadow-grass, cock's-foot and cat's-tail
Season July–June
Distribution Widely distributed in central and southern England, but is only recorded from Yorkshire in the north. It occurs in South Wales, but is not recorded from Scotland or Ireland
Notes The young caterpillars go into hibernation shortly after hatching. They are pale buff in colour and select dead brown leaves or stems of the food-plant on which to rest. As soon as the weather is mild enough, sometimes as early as January, the caterpillars become active, feeding by day on grasses. There are two colour forms of the full-grown caterpillar, one being whitish brown with brown lines and the other light green with darker lines. The full-grown caterpillars crawl into the debris at the base of the food-plant to pupate. The pupae are pale ochreous with brown speckling on the wing-cases. The butterflies emerge after about three weeks.

34 The Grayling Butterfly

Family SATYRIDAE *Hipparchia semele*

Food-plants Various grasses such as couch grass, sheep's fescue, annual meadow-grass and tufted hair-grass
Season August–June
Distribution Throughout the British Isles
Notes The caterpillars hibernate after their second moult, while still quite small. However, this hibernation is not complete as they will become active during

mild spells in winter, feeding by night or day. In spring, when the caterpillars come out of hibernation, they feed only at night, hiding at the base of the foodplant by day. The full-fed caterpillar burrows just below the surface of the ground where it forms a cell by spinning particles of earth together with silk. The reddish brown pupa is formed within this protective cell. The butterfly hatches after about a month.

35 The Gatekeeper Butterfly

Family SATYRIDAE *Pyronia tithonus*

Food-plants Grasses such as annual meadowgrass, couch grass, cock's-foot and rye grass
Season September–June
Distribution Throughout England and Wales, often very common in the south. In Ireland it is mostly confined to the southern counties. There are no recent records from Scotland
Notes The newly hatched caterpillar is grey, but after the first moult it becomes green with a brown head, feeding by day on grass blades. It goes into hibernation during October, becoming active again in April when it is greenish grey with a pale brown head. At this stage, the caterpillar feeds at night, resting at the base of the grass stems by day. When full fed it spins a pad of silk on a grass stem and pupates head downwards. The pupa is pale ochreous white with darker brown markings. The butterfly emerges after about three weeks.

36 The Meadow Brown Butterfly

Family SATYRIDAE *Maniola jurtina*

Food-plants Grasses such as meadow-grass and annual meadow-grass
Season July–June

Distribution Throughout the British Isles
Notes Although the caterpillars overwinter, they cannot be said to hibernate completely as they remain active and continue to feed except in very cold weather. They feed at night, resting low down on the grass stems during the day. The caterpillars develop very slowly, taking about ten months to become fully grown and moulting five times. The full-fed caterpillar spins a pad of silk on a grass stem and pupates hanging head downwards. The pupa is pale green with brown and black markings. The butterfly emerges after about one month.

37 The Small Heath Butterfly

Family SATYRIDAE   *Coenonympha pamphilus*

Food-plants Various grasses such as annual meadow-grass, wood meadow-grass and mat-grass
Season Throughout the year
Distribution Widespread and common throughout the British Isles
Notes The life cycle of this butterfly is complex with two broods in the year. Caterpillars of the second brood and probably some of the first brood overwinter to produce butterflies in the following spring and early summer. Thus caterpillars of the first brood produce butterflies in late summer of one year and spring of the following year, while those of the second brood produce butterflies in early summer of the following year. Overwintering caterpillars do not truly hibernate as they feed in mild weather. The pupa hangs head downwards from a pad of silk spun on a grass stem. It is rather stout and is light green at first, becoming a more brilliant green with a darker stripe down the middle of the back. The butterfly emerges after about a month.

38 The Ringlet Butterfly

Family SATYRIDAE *Aphantopus hyperantus*

Food-plants Various grasses such as annual meadow-grass, cock's-foot and couch grass usually growing in damp places

Season August–June

Distribution England, Wales and parts of Scotland as far north as Aberdeen. Locally common in Ireland

Notes The caterpillars feed slowly and go into a partial hibernation in October, continuing to feed in mild weather and becoming fully active again in the following March. They feed at night, resting in a straight position along the grass stems by day. At the slightest disturbance they will drop from the food-plant, rolling into a complete ring. The full fed caterpillar pupates at the base of the plant. The stout pupa is ochreous brown with darker brown markings. The butterfly emerges after about a fortnight.

MOTHS

39 The December Moth

Family LASIOCAMPIDAE *Poecilocampa populi*

Food-plants A wide range of trees including oak, lime, poplar, hawthorn, elm and birch. In captivity it is said to eat lettuce
Season April–June
Distribution Generally distributed in England and Wales, but less common in northern England. It occurs locally throughout Scotland and is widespread in Ireland
Notes The hairy caterpillars feed on the foliage of various trees, and when full grown may be found sunning themselves on the trunks. They pupate in June within silken cocoons spun either under loose bark on the tree or in leaf litter at the base. The pupae are reddish brown and glossy. The moths do not hatch until October.

40 The Pale Oak Eggar

Family LASIOCAMPIDAE *Trichiura crataegi*

Food-plants Various trees such as hawthorn, sloe, birch, oak and apple. Also recorded on heather and bilberry
Season April–June. In some parts of Scotland throughout the year
Distribution Throughout England, although more common in the south. Occurs locally in Wales, Ireland and Scotland
Notes The caterpillars usually feed in the evenings at the tips of the shoots, but may be found during the day basking in the sun. Although they usually become full grown and pupate at the end of June,

in some parts of Scotland they may hibernate and continue to feed in the following spring and summer to produce moths in the autumn of the second year. The full-fed caterpillars spin brownish elongate cocoons either in the leaf litter or under loose bark at the base of the tree within which they pupate. The moths usually emerge in August.

41 The Small Eggar

Family LASIOCAMPIDAE *Eriogaster lanestris*

Food-plants Hawthorn and buckthorn. Has also been found on fruit trees in gardens
Season May–July
Distribution Occurs locally throughout England, although most frequent in the south east. Local in Wales and Ireland and has also been recorded from Scotland
Notes The caterpillars are gregarious, living together within a closely woven web of silk. They leave the web to feed, but return afterwards, remaining together until almost full grown when they disperse. In July the full-fed caterpillar spins a substantial cocoon of silk which is oval in shape and pale ochreous or whitish in colour. The pupa within the cocoon is brown. Some moths emerge in the following February, but others remain in the pupa over two or three winters and have been known to emerge after as long as seven years.

42 The Lackey Moth

Family LASIOCAMPIDAE *Malacosoma neustria*

Food-plants Almost any native deciduous tree, particularly hawthorn, blackthorn, hazel and fruit trees
Season April–June
Distribution Common in southern England,

becoming scarcer further north. It is recorded from parts of Wales and is locally common in Ireland, but it is not known from Scotland

Notes The caterpillars hatch in April from eggs laid in the previous summer. They are gregarious and live in a silken web which they spin over the leaves of the food-plant. When large colonies occur on fruit trees they can cause damage by defoliation, but this is not a serious orchard pest. In fine weather, the caterpillars can be seen sunning themselves on the outside of the web, but if disturbed will drop to safety. After the final moult they disperse and when full fed spin cocoons amongst leaves of the food-plant. The oval cocoon is double layered, the inner finely woven layer of silk being coated with a yellow fluid which dries to form a powder; the outer layer of coarser silk is whitish. The pupa is black and covered with fine hairs. The moths emerge in July.

43 The Oak Eggar

Family LASIOCAMPIDAE *Lasiocampa quercus*

Food-plants Bramble, blackthorn, hawthorn, dogwood and heather. In captivity will eat a number of low-growing plants, and has been recorded feeding on ivy

Season At all times of the year according to locality

Distribution Throughout the British Isles

Notes The life cycle of this moth varies from south to north and, to some extent, from east to west. In the south the caterpillars hatch in summer and hibernate while quite small, feeding up in the following spring and pupating in June and July. Occasionally some caterpillars may become full fed in the first summer and overwinter as pupae. Further north the caterpillars overwinter while small, but take the whole of the following summer to become full grown, pupating in the autumn to produce moths in the spring of the third year. In the west of Ireland and in parts of North Wales this three-year cycle also applies. These caterpillars should be handled with extreme care as the hairs may cause severe skin irritation. The dull purplish brown pupa is enclosed in a large tough cocoon spun up within a slight web amongst leaf litter or on the food-plant.

44 The Fox Moth

Family LASIOCAMPIDAE *Macrothylacia rubi*

Food-plants Heather, bramble, bilberry, bog myrtle and burnet rose
Season July–April
Distribution Throughout the British Isles
Notes The young caterpillars are black with pale yellow bands between the segments. They feed through the summer and become full grown in October when they hibernate. In the following spring they become active again, but do not feed, and at this stage may be found basking in the sun. They pupate in March or April in cocoons spun up at the base of the food-plant or amongst grass roots. The long tubular cocoon is constructed of silk mixed with the dense hairs of the caterpillar. These hairs are highly irritant and should not be handled by those with sensitive skin. The moths emerge in May or June.

45 The Drinker Moth

Family LASIOCAMPIDAE *Philudoria potatoria*

Food-plants Various grasses such as annual meadow-grass, cock's-foot, couch grass and reed grass
Season August–June

Distribution Widespread in England and Wales. In Scotland it is mainly recorded from the west, and in Ireland it appears to be local

Notes After hatching in summer the caterpillars feed until September or October and then hibernate while still quite small. In April of the following year, they recommence feeding and become full grown in June. Showery weather appears to be ideal for these caterpillars. Their habit of drinking from drops of dew has given the moth both its common and scientific names. The brown pupa is formed within a long boat-shaped cocoon which is yellowish or brownish white. It is usually attached in an upright position to a stem of grass or reed. The moths emerge in June.

46 The Lappet Moth

Family LASIOCAMPIDAE　　　　*Gastropacha quercifolia*

Food-plants Blackthorn, hawthorn, buckthorn, sallow and sometimes apple

Season August–May

Distribution Local in England and Wales, but more frequent in south eastern England. It is not recorded from Scotland or Ireland

Notes The caterpillars hatch in summer and feed for a few weeks before hibernating in the autumn while still quite small. In spring they leave their hibernating quarters low down on the food-plant and recommence feeding on the foliage. The fleshy flaps or 'lappets' covered with long brown hairs along each side of the body of the caterpillar give rise to its common name. It becomes full grown in May, spinning a long, grey-brown, silken cocoon on the lower twigs, in which it pupates. The pupa is very dark brown covered with a whitish powder. The moths emerge in June or July.

47 The Emperor Moth

Family SATURNIIDAE *Saturnia pavonia*

Food-plants A wide range of plants, including heather, bramble, sallow and meadow-sweet
Season June–August
Distribution Widely distributed throughout the British Isles
Notes When first hatched the caterpillars are completely black. They are gregarious to begin with and feed together, but after the third moult they disperse. The full-grown caterpillars are very variable in colour, ranging from almost entirely green forms to those with heavy black banding. Each body segment bears a number of prominent wart-like structures which are usually yellow but are sometimes pink or purple. Before pupating the caterpillar spins a tough fibrous cocoon amongst the foliage of the food-plant. The moth emerges in the following spring, but some have been known to remain in the pupa stage for a second winter.

48 The Kentish Glory

Family ENDROMIDAE *Endromis versicolora*

Food-plants Birch and alder
Season May–July
Distribution Local in Scotland. At one time this moth was locally common in southern England, but there do not appear to be any recent records
Notes The young caterpillars have a characteristic habit of clustering together on twigs of the food-plant, holding on by their prolegs and bending the front part of their bodies backwards so that the thoracic legs are free. They are black at first, but soon become green with black dots. After the third moult, the black dots disappear and the coloration of the fully

developed caterpillar is assumed. The full-fed caterpillar spins a coarse-netted cocoon of silk covered with dead leaves or other debris, usually on the ground and sometimes just below the soil. The pupa is black or very dark brown. The moths usually emerge in March or April of the following year, although some remain in the pupa for two or more years.

49 The Pebble Hook-tip

Family DREPANIDAE *Drepana falcataria*

Food-plants Birch and alder
Season June–July and September–October
Distribution Widespread and sometimes common in England, Wales and Scotland. In Ireland it is local and scarce
Notes This moth is double brooded in the British Isles. The caterpillar usually lives and feeds on the underside of a leaf, turning the edges under and securing them with threads of silk. At first it is blackish with white marks, becoming green with white markings as it develops. When full fed it spins a strong brown silken cocoon to the underside of a leaf, the edges of which are also spun over with silk to form a protective shelter. The pupa is dark brown. Those of the later generation fall to the ground with the leaves where they remain through the winter. Moths emerge from overwintered pupae in May, while those from summer pupae emerge in August.

50 The Chinese Character

Family DREPANIDAE *Cilix glaucata*

Food-plants Hawthorn and blackthorn. Will sometimes eat the foliage of apple and pear

Season June–July and September–October
Distribution Widespread in England and Wales. In Scotland it is only recorded from the Lowlands, and in Ireland it is mainly recorded from the northern half of the country
Notes This moth is double brooded in the British Isles. The young caterpillars eat only the upper surfaces of the leaves, making brown blotches. Later they consume the whole leaf in the usual way, feeding rapidly and completing their development in under a month. The full-fed caterpillar pupates within a tough brown silken cocoon amongst the leaves or under loose bark. The pupa is reddish brown with greyish wing cases. Moths from overwintered pupae emerge in May, while those from summer pupae emerge in late July and early August.

51 The Peach Blossom

Family THYATIRIDAE *Thyatira batis*

Food-plants Bramble. In captivity will also eat garden raspberry
Season July–September
Distribution Widespread and sometimes common in England, Wales and Ireland. In Scotland it is local
Notes Caterpillars of this moth vary greatly in their speed of growth and development. Most become full fed and pupate in September, but others feed up rapidly and may pupate at the beginning of August. The later pupae overwinter to produce moths in the following June, but the earlier pupae sometimes produce moths in September and October of the first year. The full-fed caterpillar pupates within a silken cocoon spun up amongst leaves of the food-plant. The pupa is pale brown with darker mottling and with reddish brown wing-cases.

52 The Buff Arches

Family THYATIRIDAE *Habrosyne pyritoides*

Food-plants Bramble and wild raspberry. It is also said to feed on hawthorn and hazel
Season August–September or October
Distribution Widespread and sometimes common in England, Wales and Ireland. It does not seem to occur in Scotland
Notes The caterpillars feed on the foliage at night, retreating to a hiding place by day. Although they are usually full grown by the end of September they sometimes continue to feed into the following month. The full-fed caterpillar constructs an earthen cocoon just below the surface of the soil within which it pupates. The pupa is purplish black with reddish rings between the segments. Moths usually emerge in the following June or July, although earlier pupae may produce moths in September or October of the first year.

53 The Common Lutestring

Family THYATIRIDAE *Ochropacha duplaris*

Food-plants Usually birch, but also alder, oak and hazel
Season August–October
Distribution Widespread in the British Isles, but most common in the south and east of England
Notes The caterpillar spins leaves together to form a shelter in which it hides during the day, emerging at night to feed on the foliage. When full fed it pupates within a frail silken cocoon spun up amongst the leaves. The pupa is dull reddish brown in colour. Moths emerge in the following June. The alternative common name for this species is the Lesser Satin Moth.

54 The Frosted Green

Family THYATIRIDAE *Polyploca ridens*

Food-plant Oak
Season May–July
Distribution Widespread in England and Wales, but more frequent in the south. It does not occur in Scotland or Ireland
Notes The caterpillar hides by day in a shelter which is formed by folding over a leaf and securing it with silk. It emerges at night to feed on the foliage. When full grown it spins a silken cocoon amongst the foliage of the food-plant within which it pupates. The cocoon later falls to the ground with the dead leaves and remains there throughout the winter. Sometimes the caterpillar constructs a cocoon of dried leaves and debris at the base of the tree. The pupa is brown in colour. Moths emerge in the following April or early May.

55 The Orange Underwing

Family GEOMETRIDAE *Archiearis parthenias*

Food-plant Birch
Season April–June
Distribution Widely distributed in southern and eastern England and extends as far north as Co. Durham. Also recorded from Wales and Scotland, but does not seem to occur in Ireland
Notes The young caterpillar feeds on the catkins of birch, but later it eats the foliage. It spins leaves together to form a shelter within which it hides when not feeding. When full grown in early June it constructs a frail cocoon of silk in crevices in the bark or on the ground, within which it pupates. The pupa is brown in colour. Moths emerge in March of the following year.

56 The March Moth

Family GEOMETRIDAE *Alsophila aescularia*

Food-plants Almost all native deciduous trees and shrubs including privet, lilac, oak, hawthorn, plum, cherry, apple and rose
Season April–June
Distribution Widespread throughout the British Isles with the exception of northern Scotland
Notes This caterpillar can be distinguished from all other 'loopers' by the presence of an extra pair of prolegs on the fifth segment of the abdomen. The extra pair is much smaller than the others but is distinctly visible with the aid of a magnifying glass. This insect is a well known pest of fruit trees in orchards and gardens where the caterpillars feed on leaves, flower buds and later the young fruitlets. When full fed, they construct small ovate cocoons of silk and earth below ground in which they pupate. The pupa is brown. Moths emerge in March or sometimes as early as February. The female moths are wingless and look rather spider-like.

57 The Grass Emerald

Family GEOMETRIDAE *Pseudoterpna pruinata*

Food-plants Needle-whin, broom and furze. Will eat the foliage of laburnum in captivity
Season July–June
Distribution Moorlands and commons throughout the British Isles with the possible exception of northern Scotland
Notes The caterpillars feed at night, but remain on the food-plant by day when they rest in a rigid position strongly resembling the small green twigs. They hibernate on the food-plant while still quite small, recommencing feeding in the following spring. When

full fed they pupate within their silken cocoons spun amongst leaf litter on the surface of the soil. The pupa is a light silvery green in colour. Moths emerge in June.

58 The Large Emerald

Family GEOMETRIDAE *Geometra papilionaria*

Food-plants Birch, hazel and beech
Season July–June
Distribution Widespread throughout the British Isles with the exception of northern Scotland
Notes The newly hatched caterpillar is black but soon becomes tinged with orange, grey or brown, and by the time it is ready to hibernate has taken on the coloration of the surrounding twigs. It constructs a base of silk on a twig near to a bud and anchors itself into position for the winter. The following spring when the leaf buds are opening it becomes active again and gradually takes on a green coloration which matches the foliage on which it is feeding. When full grown it spins a delicate silken cocoon amongst dead leaves on the ground within which it pupates. The pupa is pale green dotted with pale brown on the back and suffused with brown on the wing cases. Moths hatch in June and July.

59 The Common Emerald

Family GEOMETRIDAE *Hemithea aestivaria*

Food-plants Before hibernation: low-growing herbaceous plants such as mugwort. After hibernation: oak, hawthorn, birch and other trees and shrubs
Season August–May or June
Distribution Widespread and often common in southern England and Wales, but more local in northern England. It is generally distributed and

often common in Ireland, but it is not recorded from Scotland

Notes After hatching, the caterpillar feeds on low-growing herbaceous plants until the autumn when it hibernates. In the spring it commences to feed again, but this time on the foliage of trees or shrubs. When at rest by day it holds its body out rigidly with its head pressed against its out-stretched thoracic legs. It pupates in a fragile cocoon of silk spun amongst the leaves of the food-plant. The pupa is very pale brown with darker brown wing-cases. Moths emerge in June.

60 The Little Emerald

Family GEOMETRIDAE   *Jodis lactearia*

Food-plants Birch, oak, hawthorn, sallow and other trees and shrubs
Season August–September
Distribution Widespread and sometimes common in England, Wales and Ireland. In Scotland it is locally common in southern counties
Notes The caterpillar feeds at night, remaining in a rigidly extended position on the food-plant during the day. When full fed it constructs a slight cocoon by drawing the edge of a dead leaf over with strands of silk. The pupa is bright green with bluish green wing-cases and brown eyes and cremaster. Moths emerge in the following May or June according to weather conditions.

61 The Garden Carpet

Family GEOMETRIDAE   *Xanthorhoe fluctuata*

Food-plants Many plants of the cabbage family including wallflower and horse-radish. Also said to eat the foliage of gooseberry and currant

Season June–October
Distribution Widespread throughout the British Isles
Notes This caterpillar, which is commonly found in gardens, feeds at night usually on the undersides of the leaves. When at rest it holds itself in a characteristic figure 2 shape. When full fed it goes underground and constructs a silken cocoon within which it pupates. The pupa is brown with dark orange rings between the segments. This moth is double brooded, caterpillars hatching in June producing a summer brood of moths while caterpillars hatching in late summer pupate in the autumn to produce moths in the spring.

62 The Yellow Shell

Family GEOMETRIDAE *Camptogramma bilineata*

Food-plants Chickweed, dandelion, dock, various grasses and other low-growing plants
Season August–May
Distribution Widespread and common throughout the British Isles
Notes The caterpillars vary in colour from green to various shades of brown. They feed on the foliage by night, hiding among the roots or under stones by day. In autumn the caterpillars go into hibernation while still quite small, becoming active again in the following spring. When full fed they bury themselves just below the surface of the ground, forming slight earthen cocoons in which they pupate. The pupa is dark olive on the thorax and wing-cases with the abdomen reddish brown. Moths emerge in June.

63 The Dark Spinach

Family GEOMETRIDAE *Pelurga comitata*

Food-plants Goosefoot, Good King Henry and orache
Season August–September
Distribution Widespread throughout the British Isles
Notes The caterpillar feeds on the flowers and especially the seeds of the various food-plants. It is very sluggish in habit, feeding at night and remaining on the seed vessels by day, curled into the shape of a figure 2. When full grown it pupates in a slight cocoon in the earth, usually choosing a sheltered position where the soil is loose and dry. The pupa is dark red-brown, blackened at the tip of the abdomen. Moths emerge in the following July.

64 The Common Marbled Carpet

Family GEOMETRIDAE *Chloroclysta truncata*

Food-plants Birch, sallow, willow, hawthorn, bilberry and wild strawberry. In captivity will feed readily on garden strawberry
Season June and August–March
Distribution Widespread throughout the British Isles
Notes This moth is double brooded. Caterpillars hatching in early June feed up quickly to produce a summer brood of moths while those hatching in August produce moths in the following spring. The second brood caterpillars hibernate through the winter but start to feed again in early spring. When at rest they adopt a characteristic attitude, with the head end of the body curled round in the form of a question mark. The full-fed caterpillar spins a leaf of the food-plant together with silk to form a shelter in

which it pupates. The active pupa is yellowish green and translucent.

65 The Grey Pine Carpet

Family GEOMETRIDAE *Thera obeliscata*

Food-plants Normally Scots pine, but will also eat the foliage of spruce and fir
Season September or March–early May
Distribution Throughout the British Isles where the food-plants are growing
Notes As caterpillars of this moth are recorded feeding in autumn and again in early spring it would appear that they hibernate through the winter. When at rest they take up a straight position along the needles and are very difficult to detect. The pupa is green with yellow longitudinal stripes along the abdomen. Moths emerge in late May and June. The Spruce Carpet, *Thera variata*, is a close relative which also feeds on spruce and fir but not on pine. The caterpillars are very similar, but those of the Grey Pine Carpet have pink thoracic legs whereas those of the Spruce Carpet have green legs.

66 The Winter Moth

Family GEOMETRIDAE *Operophtera brumata*

Food-plants Almost all deciduous trees and shrubs, including fruit trees. One of the few caterpillars to feed on rhododendron
Season April–May
Distribution Widespread and often common throughout the British Isles
Notes This is a well known pest of fruit trees in orchards and gardens. The caterpillar feeds at first on the young buds and later on the developed foliage, spinning several leaves together to form a shelter. When full fed, the caterpillar descends to the ground

on a silken thread and pupates just below the surface of the soil. The pupa is brown in colour. Moths emerge in late autumn or winter. The females are wingless.

67 The Twin-spot Carpet

Family GEOMETRIDAE *Perizoma didymata*

Food-plants Primrose, red campion, bilberry and coarse grasses
Season April–May
Distribution Widespread throughout the British Isles and often common
Notes The caterpillars feed up rapidly, mainly eating the flowers of the food-plants, including those of the grasses. When full grown they pupate in the ground in cocoons made of silk and earth. The pupae vary in colour from pale green to brown. Moths emerge in July.

68 The Common Pug

Family GEOMETRIDAE *Eupithecia vulgata*

Food-plants Hawthorn, sallow, bramble, bilberry, ragwort and golden rod
Season June–July and September
Distribution Widespread throughout the British Isles and often common
Notes This moth is double brooded. Caterpillars hatching in July produce moths in August while those hatching in autumn overwinter as pupae to produce moths in the following spring. The caterpillars feed up quickly and will even eat the foliage when it is withered. When full grown they pupate within earthen cocoons on the ground. The pupa is olive green with the abdomen reddish brown.

69 The Green Pug

Family GEOMETRIDAE *Chloroclystis rectangulata*

Food-plants Hawthorn, apple, pear, cherry, and sometimes apricot
Season April–May
Distribution Throughout the British Isles, often common
Notes This moth is sometimes a pest of fruit trees in orchards and gardens. The newly hatched caterpillar bores into a fruit bud and feeds inside. Later as the petals open it spins them together and feeds within the flower. When full grown it pupates either under loose bark and moss on the tree-trunk or in an earthen cocoon on the ground. The pupa is yellow tinged with olive except for the end of the abdomen which is red. The moths emerge after about a fortnight.

70 The Magpie Moth

Family GEOMETRIDAE *Abraxas grossulariata*

Food-plants Blackthorn, hawthorn, and in gardens on redcurrant, blackcurrant, gooseberry and euonymus
Season August–May or June
Distribution Widespread in the British Isles
Notes The caterpillars hibernate while quite young, usually sheltering in curled leaves amongst the food-plant, but sometimes in crevices in a nearby wall or fence. They leave their winter quarters in spring and return to the plants to feed on the young leaves. It is at this stage that they are sometimes serious pests in gardens and allotments where they may completely strip the foliage from currant and gooseberry bushes. The caterpillars feed mainly at night. When full grown they pupate within frail

transparent cocoons amongst the leaves or stems of the food-plant. The pupa is yellow at first but soon becomes a dark shining brown with yellow bands on the abdomen. Moths emerge in July.

71 The V-moth

Family GEOMETRIDAE *Semiothisa wauaria*

Food-plants Gooseberry, redcurrant, blackcurrant and flowering currant
Season April–June
Distribution Widespread in the British Isles, but very local in Ireland
Notes The caterpillar feeds at night on the younger shoots of the food-plant, resting by day on the underside of a leaf. Although it feeds on fruit bushes, showing a preference for redcurrant, it is not regarded as a pest. When disturbed it drops from the leaf and remains suspended by a silken thread until the danger is past. The full-grown caterpillar pupates in a silken web amongst leaves of the food-plant. The pupa is red-brown or blackish brown in colour. Moths emerge in July.

72 The Brimstone Moth

Family GEOMETRIDAE *Opisthograptis luteolata*

Food-plants Usually hawthorn, but also blackthorn, apple, plum, hazel and other trees
Season Throughout the year
Distribution Widespread and often common
Notes This moth has an unusual 'leap-frogging' sequence of life cycle so that moths emerge throughout the summer. Caterpillars hatching in April produce moths in August. The caterpillars from this brood overwinter and pupate in the following spring to produce moths in June. These in turn give rise to

caterpillars which pupate in the autumn to give moths in the following April. The caterpillars feed at night. When full grown they pupate in a reddish paper-like silken cocoon amongst the foliage or in leaf litter on the ground. The pupa is dark brown with the abdomen olive brown.

73 The Lilac Beauty

Family GEOMETRIDAE *Apeira syringaria*

Food-plants Privet, lilac, honeysuckle and snowberry
Season August–May or June
Distribution Throughout England and Wales, but more common in the south. Scarce in Ireland and not recorded from Scotland
Notes The caterpillars hibernate while quite small and resume feeding in the spring. They feed at night, remaining on the food-plant by day in a humped position so that they resemble dried-up leaves. When at rest they often rock from side to side, particularly if disturbed. The full-grown caterpillar pupates in a silken cocoon attached to a leaf or twig of the food-plant. The wing-cases of the pupa are dark brown and the abdomen is yellow-brown with paler mottling. Moths emerge in June and July.

74 The Early Thorn

Family GEOMETRIDAE *Selenia dentaria*

Food-plants Most deciduous trees and shrubs, including birch, alder, sallow, hawthorn and blackthorn
Season May–June and August–September
Distribution Widespread in the British Isles and often common, particularly in the south

Notes This moth is double brooded, caterpillars hatching in the spring producing moths in July, while those hatching in summer overwinter as pupae to produce moths in the following April. The caterpillars feed mainly at night, remaining on the food-plant by day when they bear a strong resemblance to thorned twigs. When full grown they pupate within tough semi-transparent cocoons reinforced with particles of earth. The pupa is dark brown with paler brown bands on the abdomen.

75 The Swallow-tailed Moth

Family GEOMETRIDAE *Ourapteryx sambucaria*

Food-plants Many trees and shrubs, including hawthorn, blackthorn, elder, privet, holly and especially ivy
Season August–June
Distribution Widespread in England, Wales and Ireland, but scarce in Scotland where it is mainly confined to the south
Notes The caterpillar feeds mainly at night, but remains on the food-plant by day. When at rest, it grips the twig with its claspers and holds the rest of its body out at an angle supported by a strong silken thread from its mouth. In this position it is almost impossible to detect. The caterpillar does not truly hibernate through the winter, as it will feed whenever the weather is warm enough. When full grown it constructs a delicate hammock-like cocoon of silk mixed with fragments of leaves, suspended from the underside of a twig. The slender pupa is brown dotted and streaked with black. Moths emerge in July.

76 The Feathered Thorn

Family GEOMETRIDAE *Colotois pennaria*

Food-plants Almost all deciduous trees and shrubs, including oak, birch, poplar, sallow, apple and hawthorn
Season April–June
Distribution Widespread throughout the British Isles, but most common in the south
Notes The caterpillar feeds mainly at night, remaining on the food-plant by day where it resembles a twig. Before pupation it constructs a large subterranean cocoon of thick silk mixed with grains of sand. The pupa is reddish brown with paler wing-cases. Moths emerge in October.

77 The Pale Brindled Beauty

Family GEOMETRIDAE *Apocheima pilosaria*

Food-plants Most deciduous trees and shrubs, including birch, oak, elm, poplar, hawthorn, blackthorn and fruit trees
Season May–June
Distribution Widespread throughout the British Isles, but scarce in Ireland
Notes The caterpillars feed at night, remaining during the day on the twigs where they are difficult to detect. When young they feed on unopened leaf buds, but later eat the foliage as it develops. The full-grown caterpillar pupates underground usually at the roots of the food-plant. At this stage there is a high mortality so that caterpillars are much more common than moths. The pupa is reddish brown in colour. Moths usually emerge in January.

78 The Brindled Beauty

Family GEOMETRIDAE *Lycia hirtaria*

Food-plants Almost all deciduous trees and shrubs, but particularly on lime, elm, willow and fruit trees such as plum and pear

Season June–July

Distribution Widespread in the British Isles, particularly common in the London area

Notes The young caterpillar is almost black with distinct yellow bands, but as it develops becomes lighter in colour, varying from reddish brown to a shade of grey. It feeds at night, resting by day on the twigs or, when almost full grown, stretched out on the trunk. When full fed it pupates below ground in a brittle earthen cocoon. The pupa is dark reddish or purplish brown. Moths emerge in the following March or April.

79 The Peppered Moth

Family GEOMETRIDAE *Biston betularia*

Food-plants Most deciduous trees and shrubs such as oak, elm, beech, sallow, fruit trees, rose and bramble

Season July–September

Distribution Widespread throughout the British Isles

Notes The caterpillars are variable in colour from shades of green to brown. They feed by night and if disturbed during the day fall to the ground where they remain for some time as though dead. The full-grown caterpillar buries itself in the soil and makes a fragile earthen cocoon within which it pupates. The pupa is dark reddish brown with the wing-cases tinged with green. Moths emerge in May or June of the following year.

80 The Mottled Umber

Family GEOMETRIDAE *Erannis defoliaria*

Food-plants Almost all deciduous trees and shrubs, including oak, birch, blackthorn, hawthorn, fruit trees, rose and honeysuckle
Season March–June
Distribution Widely distributed and often common throughout the British Isles, but less frequent in northern Scotland
Notes The caterpillars are sometimes pests of fruit trees in orchards and gardens and are also notorious for defoliating oak trees in the spring. They feed mainly at night but remain on the foliage by day. If disturbed they drop on silken threads where they remain suspended until the danger is past. They are quite variable in colour, but the basic pattern remains the same. The full-grown caterpillar descends to the ground where it pupates either on or below the surface. The pupa is reddish brown in colour. Moths emerge in October or later.

81 The Waved Umber

Family GEOMETRIDAE *Menophra abruptaria*

Food-plants Privet and lilac
Season May–August
Distribution Most frequent in the London area, although it is quite common in other parts of southern England and Wales. There are isolated records from northern England, Scotland and Ireland
Notes The caterpillars vary in colour from greyish brown to almost black and appear to be influenced by the colour of their surroundings. The full-grown caterpillar spins a tough silken cocoon on a twig of the food-plant. It gnaws a slight depression in the bark and covers the cocoon with the fragments which

form an effective camouflage. The pupa is dark red-brown in colour. Moths emerge in April of the following year.

82 The Willow Beauty

Family GEOMETRIDAE *Peribatodes rhomboidaria*

Food-plants Many trees and shrubs including hawthorn, birch, privet, lilac, rose, yew, fir and particularly ivy
Season August–May
Distribution Widespread in the British Isles
Notes The caterpillar feeds in August, but hibernates while still small on a carpet of silk spun on the lower parts of the food-plant. It becomes active again in April, feeding at night, but remaining on the food-plant by day. The full-grown caterpillar spins a tough cocoon of silk on the food-plant within which it pupates. The pupa is brown in colour. Moths emerge in July.

83 The Mottled Beauty

Family GEOMETRIDAE *Alcis repandata*

Food-plants Almost all deciduous trees and shrubs, but particularly hawthorn, birch, elm, hazel, honeysuckle, bilberry and heather
Season July–May
Distribution Widespread and common
Notes The caterpillars seem to show a preference for undergrowth rather than the foliage of larger trees. They feed mainly at night, but remain on the twigs by day. The full-grown caterpillar descends to the ground and pupates below the soil without constructing a proper cocoon. The pupa is reddish brown with the cremaster dark brown. Moths emerge in June.

84 The Pale Oak Beauty

Family GEOMETRIDAE *Serraca punctinalis*

Food-plants Oak, birch and sometimes sallow
Season July–August
Distribution Local, but sometimes common in parts of southern England. There are isolated records from Ireland, but this moth does not appear to be known from Scotland or Wales
Notes The caterpillars, which are very twig-like, vary in colour from various shades of brown to a greenish-grey. The full-fed caterpillar descends to the ground where it pupates below the surface of the soil or under moss at the base of the tree. The pupa is dark reddish brown with the cremaster blackish. Moths emerge in June of the following year.

85 The Common Heath

Family GEOMETRIDAE *Ematurga atomaria*

Food-plants Heather, heath, clover and trefoils
Season July–August
Distribution Widespread and common on heathland throughout the British Isles
Notes The caterpillars feed at night, resting on the food-plant by day in a rigidly extended posture. If disturbed they fall from the food-plant and lie as if dead. The full-grown caterpillar pupates underground without constructing a cocoon. Moths emerge in May of the following year.

86 The Bordered White

Family GEOMETRIDAE *Bupalus piniaria*

Food-plants Pine, spruce and other conifers
Season August–October
Distribution Widespread, and often common in

England, Wales and Scotland. Local, but also widespread in Ireland

Notes The caterpillars of this moth are known to the forester as 'Pine Loopers' and are often a serious pest of conifer plantations. They feed on the needles, their coloration and pattern making them very difficult to detect. The full-grown caterpillar descends to the ground where it pupates either amongst the fallen needles or below the soil. The pupa is reddish brown with the wing-cases dark greenish brown. Moths emerge in May and June, but later in the north.

87 The Early Moth

Family GEOMETRIDAE *Theria rupicapraria*

Food-plants Hawthorn, blackthorn, plum and bilberry
Season April–May
Distribution Widespread and often common throughout the British Isles with the exception of northern Scotland
Notes The usual colour of the caterpillar is light green with white markings, but varieties exist where the light green is replaced by dark green to a greater or lesser extent. It feeds up quickly in the spring and when full grown descends to the ground to pupate. The yellow-brown pupa is enclosed in a slight web on or just below the soil. Moths emerge in the following January and February.

88 The Grass Wave

Family GEOMETRIDAE *Perconia strigillaria*

Food-plants Heather, broom and the flowers of furze
Season September–April or May
Distribution Heaths and moors throughout the

British Isles, although it is more local in Scotland and Ireland

Notes The caterpillars go into hibernation while still small, sheltering at the roots of the food-plant. They recommence feeding in the spring, and when full grown pupate in a slight web, either amongst twigs of the food-plant or in debris on the ground. The pupa is shining red-brown in colour. Moths emerge in June.

89 The Death's Head Hawk-moth

Family SPHINGIDAE *Acherontia atropos*

Food-plants Potato, woody nightshade, snowberry and jasmine

Season August–September, sometimes earlier

Distribution This migrant moth is quite a rare visitor to the British Isles, although in favourable years it has been recorded from many areas. It is most frequently found in south east England

Notes There are many colour forms of the caterpillar, but those found in the British Isles are usually green or yellow with sloping blue or purple stripes along the sides of the body. It is usually found feeding on the foliage of potato, although it is quite difficult to detect when resting on the underside of a leaf. The full-fed caterpillar burrows underground and makes a large chamber within which it pupates. The large pupa is mahogany-brown in colour but turns black shortly before the emergence of the moth at which stage it can produce a squeaking sound. It is doubtful if pupae will survive the cold winter in the British Isles, but it is possible to rear moths in captivity by keeping the pupae warm indoors.

90 The Privet Hawk-moth

Family SPHINGIDAE *Sphinx ligustri*

Food-plants Privet, lilac, ash and viburnum
Season August–September
Distribution Most common in southern England, becoming scarcer further north. It is recorded from a number of areas in Wales but does not seem to occur in Scotland or Ireland
Notes The caterpillar feeds mainly at night but remains on the food-plant by day. It may often be found resting upside down on the upper part of a long spray of privet. When full fed, after about six or seven weeks, the caterpillar changes colour to a dull greenish brown before crawling away to pupate. It burrows into the soil, sometimes burying itself as deep as six inches, and excavates a pupation chamber. The large pupa is brown in colour with the tongue case prominent and separated from the rest of the body. Moths emerge in the following June or July.

91 The Lime Hawk-moth

Family SPHINGIDAE *Mimas tiliae*

Food-plants Usually lime or elm but also alder, birch and hazel
Season July–August
Distribution Common in southern England, becoming scarcer further north. Recorded also from the border counties of Wales but not from Scotland or Ireland
Notes The caterpillars feed at night resting by day on the under-surface of a leaf. When full fed they change colour to a dull pinkish brown and descend from the trees to find a suitable pupation site. They pupate either on the surface of the soil amongst leaf litter or buried about one inch below ground within

a fragile cell. The pupa is dark reddish brown and has a rough surface. Moths usually emerge in the following May, although they have been known to remain in the pupa for a second winter.

92 The Eyed Hawk-moth

Family SPHINGIDAE *Smerinthus ocellata*

Food-plants   Sallow, willow and apple
Season   Usually July–August
Distribution   Widespread in England and Wales becoming scarcer further north. It is recorded from the lowlands of Scotland and is widely distributed in Ireland
Notes   This moth is usually single brooded in the British Isles. However, in favourable seasons moths emerge in May giving rise to a second generation in July. In poor years moths of the first generation do not emerge until late June or mid July. Although not generally regarded as a pest, caterpillars have on occasions caused quite serious defoliation of young apple trees in orchards. The full-grown caterpillar descends from the tree and pupates an inch or two below the surface of the soil in an earthen chamber. The pupa is dark brown and glossy.

93 The Poplar Hawk-moth

Family SPHINGIDAE *Laothoe populi*

Food-plants   Poplar, sallow, willow and aspen
Season   July–September or October
Distribution   Widespread and often common throughout the British Isles
Notes   This moth is double brooded in favourable years. Caterpillars hatching in spring will produce moths in July which give rise to a second generation

of caterpillars in summer. These pupate in the autumn to produce moths in the following spring. In poor seasons there is only one generation, as the first moths do not emerge until July or August. The usual colour form of the caterpillar is yellowish green, but a blue green form exists which sometimes has a row of red spots along each side of the body. When full fed, the caterpillar becomes a dull greyish green and leaves the tree to pupate just below the surface of the soil. The pupa is very dark brown and rough in texture.

94 The Humming-bird Hawk-moth

Family SPHINGIDAE *Macroglossum stellatarum*

Food-plants Bedstraw. In captivity will feed on goosegrass
Season July–August
Distribution This migrant moth occurs throughout the British Isles and in favourable years is quite common
Notes The caterpillars may be found on the lower parts of the food-plant where they feed up rapidly. The young caterpillars are green, but in the final stage there are also brown forms. Just before pupation all forms become a dull purplish brown. The pale reddish brown pupa is formed within a loosely woven cocoon on the ground beneath the food-plant. Moths emerge after about three weeks. This species does not usually survive the winter in the British Isles.

95 The Elephant Hawk-moth

Family SPHINGIDAE *Deilephila elpenor*

Food-plants Willow herb, bedstraw and fuchsia
Season July–August
Distribution Throughout England and Wales, although most common in the south. It is recorded

from lowland Scotland and is widespread in Ireland
Notes The young caterpillars are basically green in colour, but most of them change to brown after the second moult although a few remain green until full grown. When disturbed the caterpillar expands the characteristic 'eye spots' on the body by drawing in its head. This gives a menacing appearance which may serve to frighten away its enemies. Although these large caterpillars may seem quite alarming they are harmless. When full grown they become darker in colour and leave the food-plant to pupate either on or just below the surface of the soil. The large pupa is pale brown with darker markings and is formed within an open cocoon of earth and debris spun together with silk. Moths emerge in the following June.

96 The Buff-tip

Family NOTODONTIDAE *Phalera bucephala*

Food-plants Most deciduous trees including elm, oak, lime, beech and hazel
Season August–September
Distribution Widespread throughout the British Isles
Notes The young caterpillars are gregarious, feeding side by side on a leaf until it is stripped. They will defoliate the upper part of a branch in this way before moving in a company to another branch to repeat the operation. When they are not feeding they cluster together on a twig where they resemble a dead leaf. They disperse when nearly full grown. The dark reddish brown pupa is formed just below the surface of the earth, often at the roots of the tree. Moths emerge in June of the following year.

PLATE 1

PLATE 2

PLATE 3

PLATE 4

PLATE 5

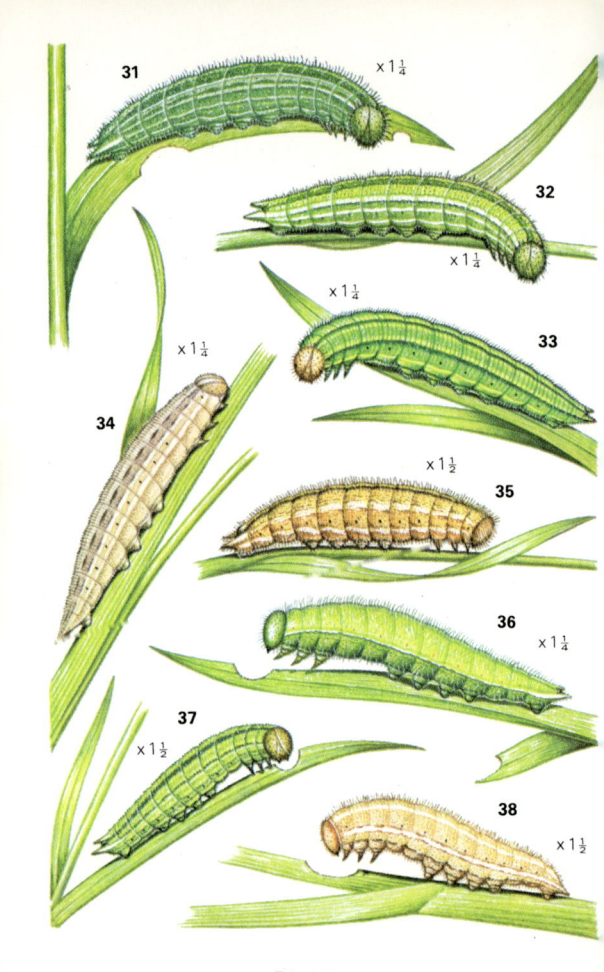

PLATE 6

PLATE 7

PLATE 8

PLATE 9

PLATE 10 all x 1½

PLATE 11 all × 1½

PLATE 12

PLATE 13

PLATE 14 all × 1¼

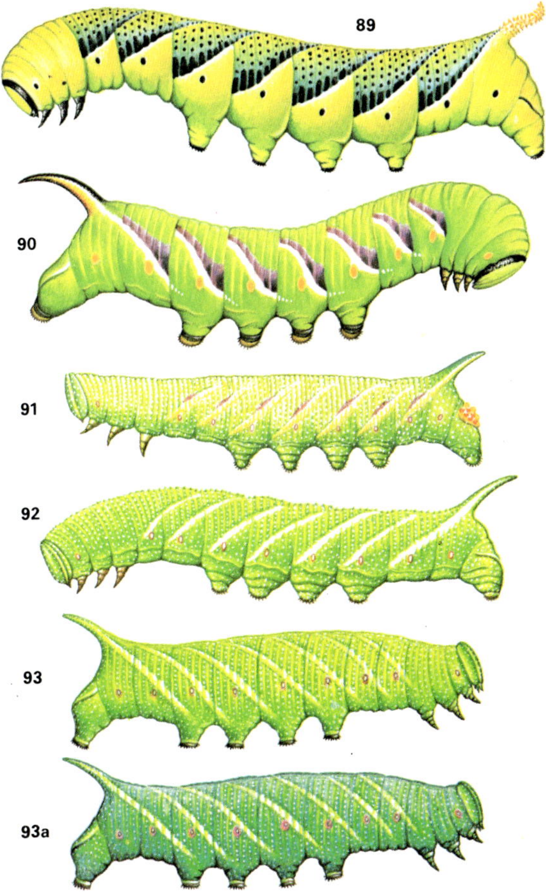

PLATE 15

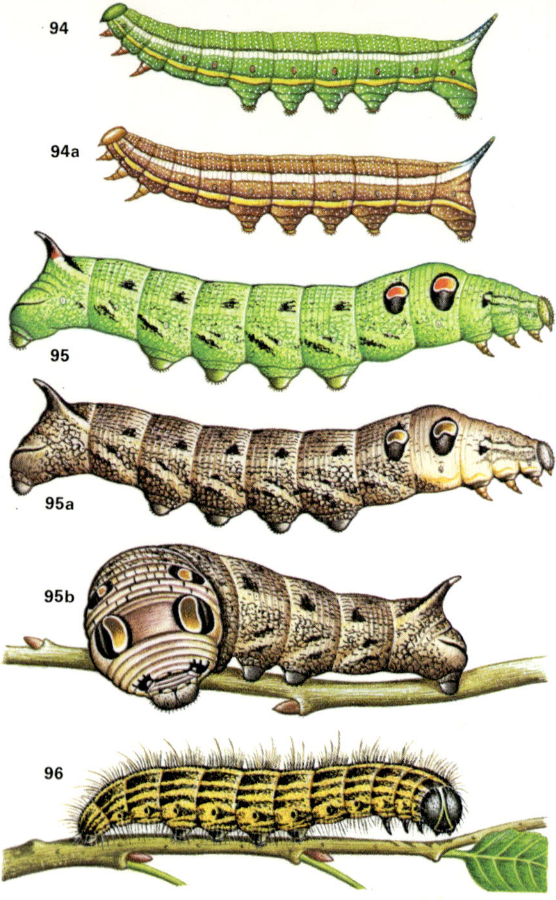

PLATE 16

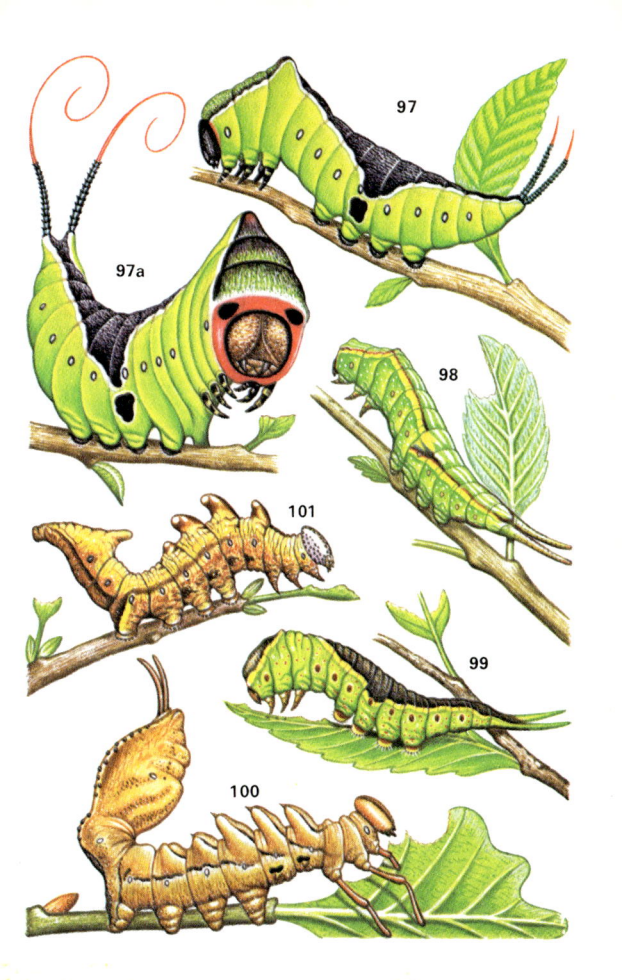

PLATE 17

PLATE 18

PLATE 19

PLATE 20

PLATE 21

PLATE 22

PLATE 23

PLATE 24

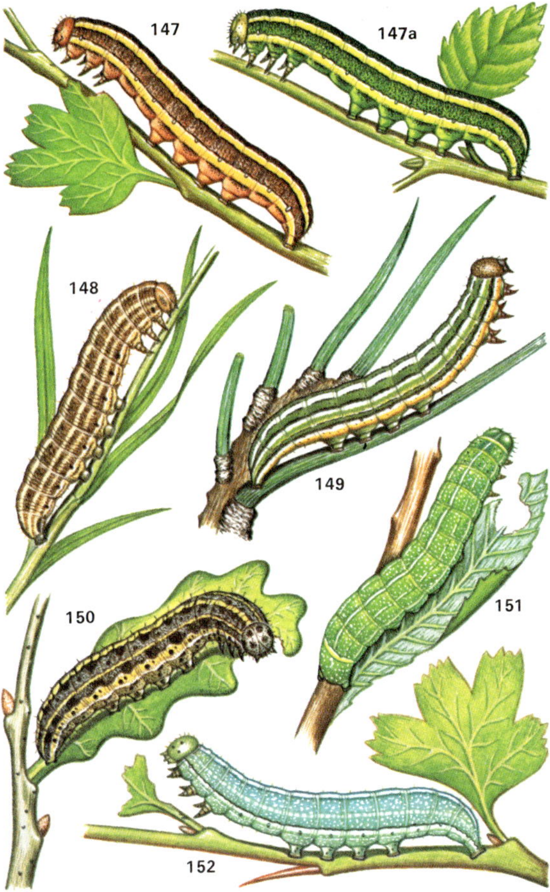

PLATE 25   all x 1¼

PLATE 26

all × 1¼

PLATE 27   all x 1¼

PLATE 28

PLATE 29　　all x 1½

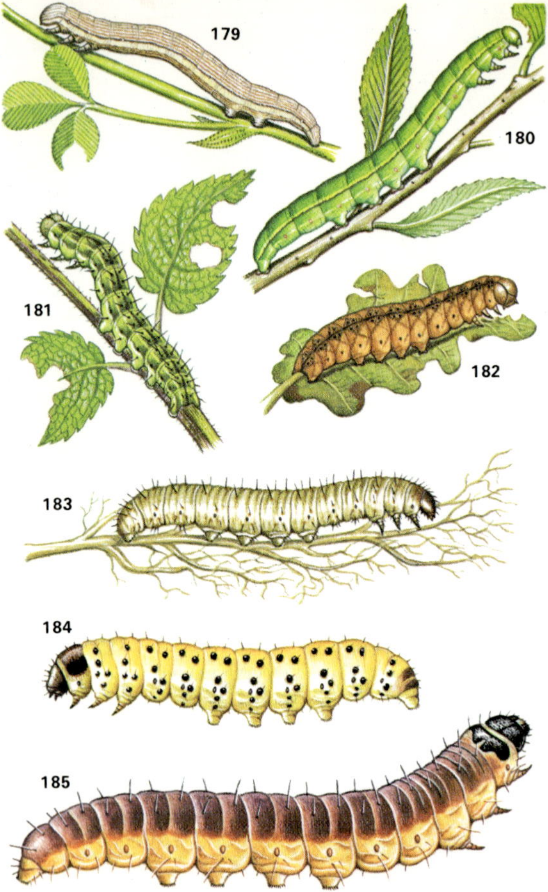

PLATE 30 all × 1¼

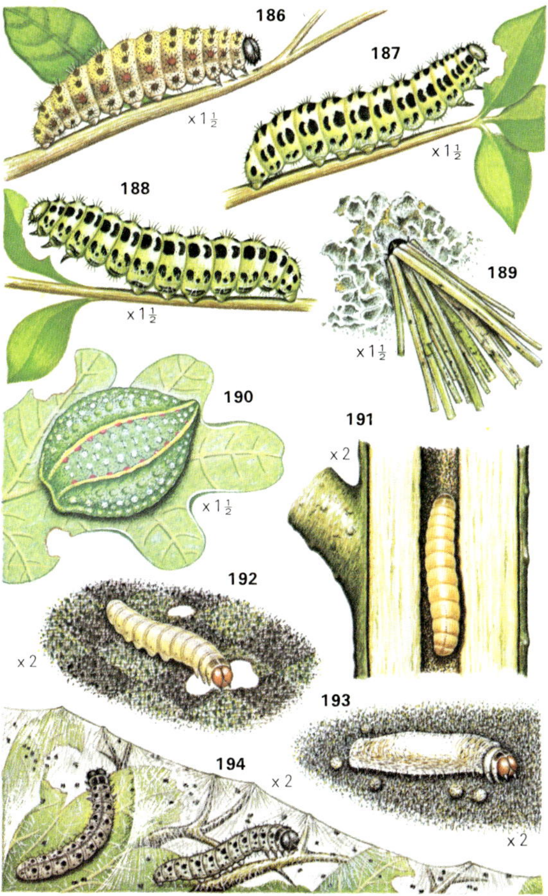

PLATE 31

PLATE 32 all x 2

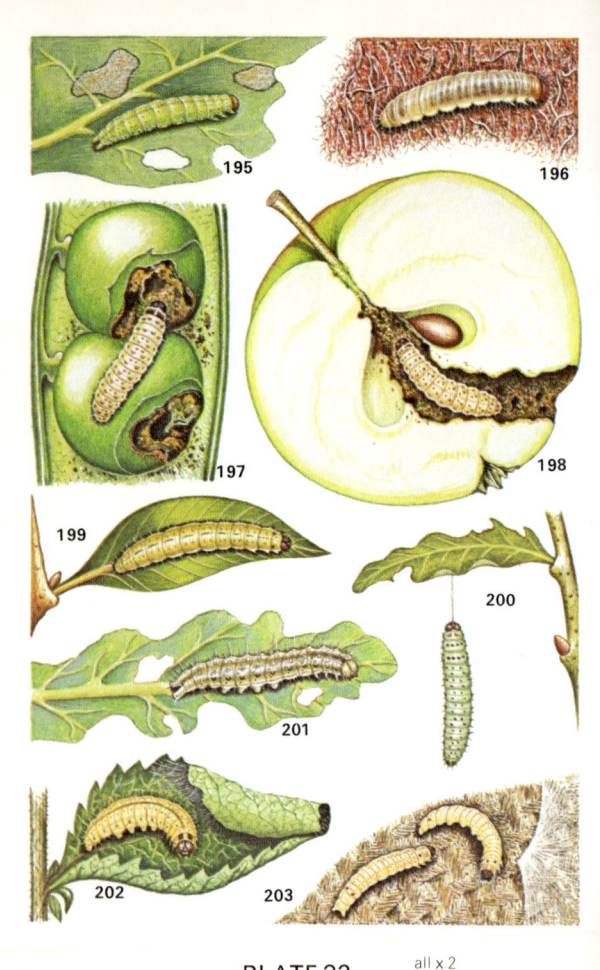

97 The Puss Moth

Family NOTODONTIDAE *Cerura vinula*

Food-plants Sallow, willow and poplar
Season July–September
Distribution Widespread in the British Isles
Notes The caterpillar often sits conspicuously on the upper surface of a leaf, feeding in the daytime as well as at night. If disturbed it will assume a fierce defensive attitude with its front parts lifted up and the characteristic tails raised and curved forward over the back. When in this position, bright red filaments called flagellae are extruded from the whip-like tails to add to the threatening effect. The young caterpillar has a pair of ear-like projections on the front of the body just behind the head, but these disappear as it grows. When full fed it forms a tough shell-like cocoon strengthened with fragments of bark. The cocoon is usually found on a tree-trunk about four feet from the ground. The pupa is purplish brown with darker wing-cases. Moths generally emerge in the following May, but sometimes spend a second winter in the pupa stage.

98 The Sallow Kitten

Family NOTODONTIDAE *Harpyia furcula*

Food-plants Sallow and willow, sometimes aspen
Season July–September
Distribution Widespread throughout the British Isles
Notes The full-fed caterpillar makes a tough cocoon of silk strengthened with fragments of wood and bark which is attached to the trunk of a tree. It usually selects a depression in the bark where the cocoon is very difficult to detect. The pupa is dark

reddish brown with the wing-cases tinged with green. Moths usually emerge in May.

99 The Poplar Kitten

Family NOTODONTIDAE *Harpyia bifida*

Food-plants Poplar and aspen
Season July–September
Distribution Widespread in England, but less common in the north. Recorded from parts of Wales, but rare in Ireland and absent from Scotland
Notes The full-grown caterpillar constructs a rigid, tough cocoon from silk mixed with fragments of gnawed wood and bark. The cocoon is sometimes spun on a small branch, but is more usually found in a crevice on the tree-trunk one to three feet above the ground. The pupa is reddish brown in colour. Moths usually emerge in June of the following year, but sometimes spend a second winter in the pupa.

100 The Lobster Moth

Family NOTODONTIDAE *Stauropus fagi*

Food-plants Beech, birch, oak, hazel and apple
Season July–September
Distribution Widely distributed, but local in Wales and the southern half of England, while in Ireland it is extremely local and confined to the south west. There are no records from Scotland
Notes The newly hatched caterpillar looks very much like an ant, and is also remarkable for the fact that it eats nothing but its egg-shell during the first week. As it develops it takes on the characteristic 'lobster' appearance which gives the moth its name. When full grown the caterpillar spins a tightly woven silken cocoon between fallen leaves within which it pupates. The pupa is blackish-brown with a violet bloom. Moths usually emerge in the following May.

101 The Iron Prominent

Family NOTODONTIDAE *Notodonta dromedarius*

Food-plants Birch, alder and hazel
Season June–August and sometimes September–October
Distribution Widespread throughout the British Isles
Notes Although generally single brooded this moth may produce a second generation in certain localities when conditions are favourable. The caterpillar is most frequently found on isolated and badly grown trees where it feeds at the tips of the lower branches. Before pupation it descends to the foot of the tree where it spins a cocoon just below the surface of the soil. The tough cocoon is fairly large and is constructed of silk mixed with loose earth. The pupa is glossy and dark brown in colour. Moths emerge in June from pupae of the previous year and in August as a second generation.

102 The Lesser Swallow Prominent

Family NOTODONTIDAE *Pheosia gnoma*

Food-plant Birch
Season June–July and sometimes September–October
Distribution Widespread throughout the British Isles, but tends to be more common in the north of England and Scotland
Notes This moth is generally single brooded, although there is sometimes a second generation. The caterpillars, which have a shining varnished appearance, rest on the underside of the leaves, often on the leaf stalk. In captivity they tend to be cannibalistic and so should be isolated. When full grown they pupate underground within tough cocoons of silk

mixed with earth. The elongated pupa is dark purplish brown. Moths from overwintering pupae emerge in May and those of the second generation in July.

103 The Swallow Prominent

Family NOTODONTIDAE *Pheosia tremula*

Food-plants Poplar and sallow
Season June–July and September–October
Distribution Widespread in the British Isles, but most common in south eastern England
Notes This moth is double brooded, caterpillars hatching in July producing moths in August while those hatching in autumn overwinter as pupae to produce moths in the following spring and early summer. The caterpillars seem to be difficult to rear in captivity and have a high moisture requirement. They will readily drink from drops of water. In autumn, the full-grown caterpillar constructs a subterranean cocoon of coarse grey silk mixed with earth, often at the foot of the tree. Summer cocoons are, however, spun up between leaves of the food-plant. The pupa is deep reddish brown with the thorax and wing-cases darker.

104 The Coxcomb Prominent

Family NOTODONTIDAE *Ptilodon capucina*

Food-plants Various trees and bushes, but mainly birch, oak, hazel, sallow and beech
Season June–July and September–October
Distribution Widespread throughout the British Isles
Notes This moth is double brooded in southern England but single brooded in the north. Before pupation the caterpillar constructs a fine silken cocoon covered with particles of earth usually

attached to a stone or to a root underground. The pupa is dark purplish brown in colour. Moths of the first brood hatch in May and those of the second brood in July and August.

105 The Small Chocolate-tip

Family NOTODONTIDAE *Clostera pigra*

Food-plants Sallow and aspen
Season June–September
Distribution Widespread in the British Isles
Notes This moth is double brooded and may also produce a third generation. Caterpillars may be found in various stages of development throughout the summer and autumn. Moths from overwintered pupae emerge in May and those of the second generation in July and August, and again in October. The caterpillar usually feeds on low-growing species of sallow. It hides by day in a shelter, formed by drawing together a few leaves with silk, and emerges at night to feed. When full grown it pupates within a loosely formed cocoon of silk among fallen leaves and debris on the ground or sometimes between leaves on the bush. The pupa is bright reddish brown in colour.

106 The Figure of Eight Moth

Family NOTODONTIDAE *Diloba caeruleocephala*

Food-plants Hawthorn, blackthorn and fruit trees such as apple and plum
Season April–June
Distribution Widely distributed in England, Wales and Ireland, but most common in south east England. It has also been recorded from lowland Scotland
Notes The caterpillars feed quite exposed on the leaves, showing a preference for the end shoots of the

lower branches. They can sometimes be seen in large numbers feeding on hedges. The full-grown caterpillar spins a strong cocoon of silk mixed with fragments of leaf and bark or particles of soil. It is formed either on the ground, under bark or attached to a twig at the base of a hedge. The pupa is dull reddish brown with a bluish powdering. Moths emerge in October and November.

107 The Vapourer Moth

Family LYMANTRIIDAE *Orgyia antiqua*

Food-plants Most deciduous trees and shrubs, including oak, lime, blackthorn, hawthorn, fruit trees and rose
Season May–August
Distribution Widespread throughout the British Isles but less common in Ireland
Notes This moth is particularly common in the London area where the caterpillars sometimes occur in vast numbers on trees in parks and gardens. Care should be taken in handling them as they possess irritant hairs which may cause rashes on sensitive skins. The eggs hatch out over a long period, so that caterpillars may be found at various stages of development throughout the summer. When full fed the caterpillar spins a very thin but tough cocoon of silk interwoven with hairs from its body. Cocoons may be found in crevices in bark, on fences and walls, but those of the females are commonly spun up amongst twigs of the food-plant. The pupa is a glossy brownish black with many small tufts of short whitish hairs. Moths emerge in June of the following year. Females are wingless.

108 The Dark Tussock

Family LYMANTRIIDAE *Dasychira fascelina*

Food-plants Hawthorn, sallow, willow, broom and ling
Season August–April or May
Distribution Northern England, North Wales, Scotland and also parts of southern England. In Ireland it is very local in the northern half
Notes The caterpillars hibernate when still small in a thin silken web in the fork of a branch of the food-plant. Three or four individuals share a common nest which is protected by spinning a few dead leaves together to form a shelter. They become active again in spring and continue to feed, but are difficult to rear in captivity unless kept on a growing plant. The full-grown caterpillar pupates within a tough cocoon of silk interwoven with hairs from its body. Cocoons are usually found amongst leaves of the food-plant close to the ground. The shining black pupa is densely covered with tufts of brown hairs. Moths emerge in June or July.

109 The Pale Tussock

Family LYMANTRIIDAE *Dasychira pudibunda*

Food-plants Various deciduous trees and shrubs such as oak, birch, hazel and wild and cultivated hop
Season July–September
Distribution Throughout England and Wales, but more common in the south. It is widespread in the southern half of Ireland but does not appear to occur in Scotland
Notes The caterpillars feed openly on the foliage where they remain when at rest. At one time they

were very abundant in hop gardens where they were known to the pickers as 'hop dogs'. Fortunately they feed on the foliage too late in the year to cause serious damage to this crop. Pupation takes place within a large silken cocoon spun up amongst the leaves close to the ground. The pupa is chestnut brown in colour and covered with hairs. Moths emerge in May of the following year.

110 The Brown-tail Moth

Family LYMANTRIIDAE *Euproctis chrysorrhoea*

Food-plants Hawthorn, blackthorn, sea buckthorn, rose and fruit trees such as plum and apple
Season August–June
Distribution Mainly confined to south east England where it is sometimes very common in coastal areas
Notes The young caterpillars construct a communal nest when very small in which they hibernate through the winter. When they recommence feeding in the spring further nests are constructed as the caterpillars grow. They leave the nest to feed, but return to rest and to moult. When in large numbers they can cause serious defoliation of hedges and fruit trees, and are difficult to control by spraying as they are protected by their body hairs. These hairs are highly irritant and can cause serious inflammation of sensitive areas such as the face. The full-grown caterpillar pupates within a strong brown cocoon of silk interwoven with body hairs. Sometimes five or six cocoons are enclosed in a common web spun on the food-plant. The pupa is blackish brown with tufts of hairs on the body. Moths emerge in July and August.

111 The Yellow-tail Moth

Family LYMANTRIIDAE *Euproctis similis*

Food-plants Usually hawthorn, but also oak, beech, birch, sallow, rose and fruit trees such as apple and pear
Season August or September–May or June
Distribution Widespread in England and Wales. There are isolated records from Scotland and Ireland
Notes The young caterpillars protect their communal web by surrounding it with leaves which are spun together. They only leave the nest to feed, returning afterwards. In October each caterpillar spins a silken case within which it hibernates through the winter. They recommence feeding in spring, gnawing the unopened buds until the leaves appear and living gregariously until the last moult when they disperse. At this stage they feed openly and do not return to the nest. The hairs of this caterpillar can cause irritation to sensitive skins, although not as severe as that caused by the Brown-tail Moth. Pupation takes place within a brownish silken cocoon spun up amongst leaves or twigs of the food-plant. The pupa is dark brown and slightly hairy. Moths emerge in June and July.

112 The White Satin

Family LYMANTRIIDAE *Leucoma salicis*

Food-plants Poplar, willow and sallow
Season August–June
Distribution Widespread in England, but most common in the south east and north west. There are isolated records from Wales, Scotland and Ireland
Notes The young caterpillars hibernate when not more than a quarter of an inch long, each within a small cocoon-like web in a crevice of the bark. In

April they become active again feeding mainly at night and hiding by day on the undersides of the branches. When full grown the caterpillars pupate within flimsy brownish cocoons spun up amongst leaves of the food-plant. The pupa is a glossy black with tufts of yellowish white hairs. Moths emerge in July and August.

113 The Black Arches

Family LYMANTRIIDAE *Lymantria monacha*

Food-plants Oak, birch, elm, beech, apple, Scots pine and various other trees
Season April–July
Distribution Throughout southern England and also in parts of Wales. It does not occur in Scotland and is very rare in Ireland
Notes After hatching the young caterpillars are said to rest on the tree-trunk for a few days before commencing to feed, but after this make rapid growth. They feed at night but return to the tree-trunk to rest by day. Although caterpillars of this moth are sometimes a serious pest on the Continent they do little damage in the British Isles as they feed only on mature leaves and do not injure the young shoots. Pupation takes place within a semi-transparent cocoon of white silk spun up in a crevice of the bark, frequently on the trunk. The pupa is a polished dark brown with numerous tufts of hairs. Moths emerge in July and August.

114 The Muslin Footman

Family ARCTIIDAE *Nudaria mundana*

Food-plants Lichens and algae
Season August–May or June
Distribution Widespread throughout the British Isles

Notes The caterpillars feed especially on lichens growing on stone walls. Although they sometimes feed in sunshine they usually hide under stones by day and feed at night. They hibernate through the winter becoming active again in spring. The full-grown caterpillar pupates within a semi-transparent cocoon in a sheltered place among stones or on a wall. The pupa is pale green or yellowish green in colour. Moths emerge in July.

115 The Dingy Footman

Family Arctiidae *Eilema griseola*

Food-plant Lichens. In captivity will eat wilted leaves of lettuce
Season August–June
Distribution Throughout Wales and the southern half of England. It does not occur in Scotland or Ireland
Notes The caterpillars generally feed on lichens growing on trees and bushes in swampy places. They hibernate through the winter and are most frequently found when they recommence feeding in the spring. The full-grown caterpillar pupates within a thin silken cocoon interwoven with fragments of moss and lichen. The pupa is reddish brown in colour. Moths emerge in July.

116 The Common Footman

Family Arctiidae *Eilema lurideola*

Food-plants Various trees such as oak, sallow, blackthorn and buckthorn
Season August–June
Distribution Widespread throughout the British Isles
Notes Although at one time the caterpillars were believed to feed solely on lichens and algae growing

on trees it now appears that the leaves of trees may be the more usual food. After hibernating through the winter the caterpillars recommence feeding in the spring, becoming full grown in June. Pupation takes place within a thin silken cocoon spun in a crevice of the bark. The pupa is chestnut brown in colour. Moths usually emerge in July.

117 The Garden Tiger

Family ARCTIIDAE *Arctia caja*

Food-plants Herbaceous plants, particularly dead-nettles and dock, and also the foliage of trees and shrubs
Season September–June
Distribution Widespread and often common throughout the British Isles
Notes The hairy caterpillars of this moth are known as 'Woolly Bears' and are commonly found in gardens where they will feed on a wide range of weeds and cultivated plants. The newly hatched caterpillars are brownish orange with black spots but become darker as they develop. About three weeks after hatching they go into hibernation and do not become active again until the first warm days of spring when they recommence feeding and may also be found basking in the sun. In captivity, some caterpillars will feed up rapidly and pupate in autumn of the same year. The undersides of dock leaves are a favourite resting place. It should be noted that the hairs of these caterpillars can cause a rash when handled by those with sensitive skins. The full-fed caterpillar spins a dull yellowish cocoon of loosely woven silk amongst low herbage or in rubbish on the ground. The pupa is yellowish brown at first, later becoming a dark purple brown or black. Moths emerge in July.

118 The Cream-Spot Tiger

Family ARCTIIDAE *Arctia villica*

Food-plants Many low-growing plants, particularly dandelion, chickweed, groundsel and dock
Season July–April
Distribution Widespread in southern England and also recorded from South Wales
Notes The young caterpillars go into hibernation a few weeks after hatching and while still small. They become active again in March and feed up rapidly. When full grown they pupate in large loosely spun silken cocoons amongst foliage of the food-plants or on the ground. The pupa is black. Moths emerge in May and June.

119 The Clouded Buff

Family ARCTIIDAE *Diacrisia sannio*

Food-plants Low-growing plants such as dandelion, dock, chickweed and plantain
Season July–April or May
Distribution Heaths and mosses throughout the British Isles
Notes The caterpillars usually hibernate when quite small, although in captivity some may become full fed in August. They recommence feeding in the spring, becoming full grown in April or May. These caterpillars are very active but coil up when disturbed. When young, the spots on the dorsal stripe are crimson, but these become yellow as the caterpillar grows. Pupation takes place within a frail silken cocoon spun up at the roots of the plants. The pupa is dark brown with pale grey streaks on the head and wing-cases. Moths emerge in June.

120 The White Ermine

Family ARCTIIDAE *Spilosoma lubricipeda*

Food-plants Most low-growing plants
Season August–September
Distribution Widespread throughout the British Isles and often common
Notes The caterpillars of this moth are frequently found in gardens where they feed on weeds and cultivated plants. They are very active and when disturbed will run off at great speed, thus giving rise to the scientific name of this species which means 'fleet-footed'. The full-grown caterpillar pupates within a close-fitting cocoon of silk interwoven with body hairs. Cocoons may be spun up in a folded leaf or under any shelter on the ground, particularly at the base of a wall or fence. The pupa is a shining dark red-brown. Moths emerge in June of the following year.

121 The Buff Ermine

Family ARCTIIDAE *Spilosoma luteum*

Food-plants Many low-growing plants such as dock and plantain. Also on birch and virginia creeper
Season August–October
Distribution Widespread and often common in England, Wales and Ireland. In Scotland it is local, most records coming from the north west
Notes When very young, the caterpillars are a uniform pale yellow, but the colour changes through grey to brown as they grow. They are commonly found in gardens feeding on weeds and cultivated plants. When full grown they pupate within tightly fitting cocoons of silk interwoven with body hairs. Cocoons may be found on the ground under leaf litter or any available shelter. The pupa is dark red-brown and glossy. Moths emerge in June of the following

year, although in captivity they sometimes emerge in autumn of the first year.

122 The Muslin Moth

Family ARCTIIDAE *Diaphora mendica*

Food-plants A wide range of low-growing plants such as dandelion, dock and chickweed. Also birch and rose
Season June or July–August
Distribution Widespread and often common in England, Wales and Ireland. There appear to be very few records from Scotland
Notes The newly hatched caterpillars are yellowish white and semi-transparent but become greyish after the first moult. If disturbed they will drop from the food-plant and curl in a ring but, after a short time, they uncurl and rush away. The full-grown caterpillar pupates within a thin, close-fitting cocoon of silk mixed with body hairs and particles of earth. Cocoons may often be found in leaf litter on weedy banks or under any suitable shelter. The pupa is very dark brown and glossy. Moths emerge in May or June of the following year.

123 The Ruby Tiger

Family ARCTIIDAE *Phragmatobia fuliginosa*

Food-plants Low-growing plants such as dandelion, plantain, dock and golden rod
Season June–August and September–May
Distribution Widespread throughout the British Isles
Notes This moth is usually double brooded, although under favourable conditions a third generation may be produced. Caterpillars of the second generation become full grown in the autumn before hibernating at the roots of the food-plants. They

become active again in February or March and may be seen sunning themselves before spinning up. The fairly large cocoons are spun amongst low vegetation near the ground. The pupa is black marked with yellow on the hind edge of each segment. Moths from overwintered caterpillars emerge in May and June, while those of the summer brood emerge in July and August.

124 The Scarlet Tiger

Family ARCTIIDAE *Callimorpha dominula*

Food-plants A wide range of low-growing plants and bushes such as nettle, groundsel, bramble, sallow and blackthorn
Season July or August–April or May
Distribution Restricted to the southern half of England and Wales
Notes The young caterpillars hibernate while still quite small, sheltering amongst the food-plant close to the ground. They emerge in April and feed up rapidly. At this stage they may be seen feeding gregariously in full sunshine. Pupation takes place within a frail cocoon of white silk spun amongst leaves and rubbish on the ground. The pupa is dark reddish brown shaded with black. Moths emerge in June.

125 The Cinnabar Moth

Family ARCTIIDAE *Tyria jacobaeae*

Food-plants Ragwort, groundsel and coltsfoot
Season June–August
Distribution Widespread and often common throughout the British Isles, although it is more local in Scotland
Notes The caterpillars often feed gregariously in the daytime, sometimes in such numbers that whole areas of the food-plant may be completely stripped,

leaving only bare stumps. When full grown they pupate just below the surface of the soil or under moss or leaf litter, each in a frail silken cocoon. The pupa is dark reddish brown in colour. Moths emerge in the following May or June.

126 The Short-cloaked Moth

Family NOLIDAE *Nola cucullatella*

Food-plants Blackthorn, hawthorn, apple, pear and plum
Season August–June
Distribution Widespread in England and Wales, but more common in the south. Not recorded from Scotland or Ireland
Notes After hatching in August the caterpillar feeds for a while before hibernating in mid-September. At this stage the small caterpillar hides in a crevice or depression in the bark which it spins over with a few strands of silk. In May it comes out of hibernation and feeds openly on the upper sides of leaves by day and night. When full grown it pupates within a tight-fitting, spindle-shaped cocoon of whitish silk covered with fragments of bark. The cocoon is attached to a twig or small branch and is very difficult to detect. The pupa is dull brown in colour. Moths emerge in June or July.

127 The Garden Dart

Family NOCTUIDAE *Euxoa nigricans*

Food-plants Clover, plantain, dock and various other low-growing plants
Season September–June
Distribution Throughout the British Isles
Notes Caterpillars of this moth can be a serious pest in clover fields and are sometimes a nuisance in gardens. They are 'cutworms', feeding at the base of

the young plants and causing them to wilt and die. During the winter months they remain underground in cold weather, but become active in mild spells. The full-grown caterpillar pupates just below the ground in a cocoon of silk mixed with particles of soil. The pupa is light brown in colour. Moths emerge in July and August.

128 The Turnip Moth

Family NOCTUIDAE *Agrotis segetum*

Food-plants A very wide range of low-growing plants, including cultivated crops such as turnip, carrot and cabbage
Season July–August
Distribution Widespread and often very common throughout the British Isles
Notes Caterpillars of this moth are frequently a serious pest of farm crops. They hide in the ground by day coming to the surface at night to feed at the base of young plants. These 'cutworms' also cause damage by eating large cavities in roots of turnips and swedes, especially in dry seasons. The caterpillars become full fed in October and hibernate in this stage before pupating in the spring. The reddish brown pupa is formed beneath the soil in a small earthen chamber. Moths emerge in June.

129 The Shuttle-shaped Dart

Family NOCTUIDAE *Agrotis puta*

Food-plants Low-growing plants such as dandelion, dock and knotgrass
Season September–April
Distribution Widespread in southern England and Wales, becoming less common further north. It is scarce in Scotland and has only been recorded once from Ireland

Notes Although some caterpillars of this moth may appear to be full fed in December they do not pupate until the following May. Others feed throughout the winter when conditions are warm enough and also pupate in the spring. The reddish brown pupa is formed beneath the soil, usually near the base of the food-plant. Moths emerge in July.

130 The Flame Moth

Family NOCTUIDAE *Axylia putris*

Food-plants Low-growing plants such as bedstraw, dock, plantain and knotgrass
Season July–October
Distribution Widespread throughout the British Isles and often common
Notes The caterpillars generally feed at night, hiding under leaves of the food-plant by day. When full grown they pupate within fragile earthen cocoons beneath the ground at the roots of the food-plant. The pupa is dull reddish brown shaded with grey on the back. Moths emerge in June or July of the following year.

131 The Large Yellow Underwing

Family NOCTUIDAE *Noctua pronuba*

Food-plants A very wide range of low-growing wild and cultivated plants
Season August–May
Distribution Widespread and common throughout the British Isles
Notes The caterpillars of this moth can be a serious pest of field crops and are also common in gardens where they feed on weeds, vegetables and flowers. In habit they are 'cutworms', hiding underground by day and emerging at night to feed at the surface of the soil. They feed throughout the winter when the

weather is mild enough. The full-grown caterpillar is usually brown or greyish, but green forms also occur. Pupation takes place below ground within a very fragile earthen chamber. The pupa is shining reddish brown with darker wing-cases. Moths usually emerge in June or July.

132 The Lesser Yellow Underwing

Family NOCTUIDAE *Noctua comes*

Food-plants Dock, plantain, clover and other low-growing plants. After hibernation, hawthorn, blackthorn and sallow
Season September–April
Distribution Widespread and often common throughout the British Isles
Notes The small caterpillars feed on low-growing plants before hibernation, but in the spring climb the stems of bushes to feed on the opening buds and young leaves. They feed only at night, hiding amongst rubbish on the ground or under the lower leaves of the food-plant by day. Pupation takes place within a fragile earthen cocoon underground. The pupa is a shining reddish brown. Moths emerge in July.

133 The Broad-bordered Yellow Underwing

Family NOCTUIDAE *Noctua fimbriata*

Food-plants Primrose, violet, dock and other low plants. After hibernation, hawthorn, blackthorn, sallow, birch and other trees and shrubs
Season September–May
Distribution Widespread in woodlands throughout the British Isles
Notes The young caterpillars feed on low-growing plants in the autumn, but, after hibernation, feed almost exclusively on the expanding buds and young

leaves of trees and shrubs. They hide by day in leaf litter on the ground, climbing up each night to feed on the foliage. When full fed, pupation takes place within a very frail earthen cocoon just below the surface of the soil. The pupa is dark reddish brown and shining. Moths emerge in June.

134 The True Lover's Knot

Family NOCTUIDAE *Lycophotia porphyrea*

Food-plant Heather
Season August–May
Distribution Heaths and moorland throughout the British Isles
Notes After feeding in late summer the small caterpillars hibernate in the autumn. They become active again in April and recommence feeding on the foliage. Feeding takes place by night, the caterpillars hiding by day under moss and leaf litter under the food-plant. When full grown they pupate within frail silken cocoons on or just below the surface of the ground. The pupa is very dark reddish brown and glossy. Moths emerge in June or July.

135 The Ingrailed Clay

Family NOCTUIDAE *Diarsia mendica*

Food-plants Primrose, dock and other low-growing plants. After hibernation, bramble, hawthorn, blackthorn and sallow
Season August–May
Distribution Widespread throughout the British Isles
Notes The young caterpillars feed on low-growing plants until October when they go into hibernation. They become active again in the spring, feeding on the young foliage of trees and shrubs. The full-grown caterpillar pupates within a loose cocoon of earth and

silk beneath the surface of the soil. The pupa is reddish brown and very glossy. Moths emerge in June, although caterpillars reared in captivity may feed up quickly to produce moths in the autumn of the first year.

136 The Purple Clay

Family NOCTUIDAE *Diarsia brunnea*

Food-plants Dock, wood-rush, bilberry and other low-growing plants. After hibernation, bramble, sallow and birch
Season September–May
Distribution Widespread in woodland throughout the British Isles, and often common
Notes Before hibernation the caterpillars feed on various low-growing plants, but in spring they feed on the opening buds of trees and shrubs. They feed only at night, hiding under leaf litter on the ground by day. Pupation takes place within a fragile cocoon of earth and silk beneath the surface of the soil. The pupa is dark reddish brown and very glossy. Moths emerge in June.

137 Setaceous Hebrew Character

Family NOCTUIDAE *Xestia c-nigrum*

Food-plants Chickweed, plantain, groundsel, dock and many other low-growing plants
Season September–May
Distribution Widespread throughout the British Isles and often common
Notes The caterpillars are sometimes common in gardens where they will feed on both weeds and cultivated plants. They usually feed throughout the winter, except in very cold weather, and pupate in the spring, although some feed up rapidly and pupate in the autumn. The shining reddish brown pupa is

formed below the surface of the soil. Moths from spring pupae usually emerge in August, while those from autumn pupae emerge in May or June of the following year.

138 The Square-spot Rustic

Family NOCTUIDAE *Xestia xanthographa*

Food-plants Dock, chickweed, plantain, grasses and, in spring, the young shoots of sallow and hawthorn
Season August–May
Distribution Widespread and common throughout the British Isles
Notes After hatching, the young caterpillars feed for a time on low-growing plants before hibernating amongst the roots. In spring they resume feeding, and at this stage will also eat the developing foliage of bushes. Both before and after hibernation they feed at night, hiding by day at the base of the food-plant or under stones. Pupation takes place within a cocoon just below the surface of the ground. After spinning the cocoon, the caterpillar remains inside for as long as two months before changing into a pupa. The pupa is a shining reddish brown. Moths emerge in August.

139 The Heath Rustic

Family NOCTUIDAE *Xestia agathina*

Food-plant Heather
Season September–June
Distribution Heaths and moorlands throughout the British Isles
Notes The caterpillars feed through the winter except in very cold weather. They feed at night, but remain on the food-plant by day clinging closely to the stems. The colour pattern blends with the dead

and living twigs of the food-plant making it very difficult to detect. When full fed the caterpillar spins a silken cocoon low down amongst the food-plant within which it pupates. The pupa is brown and glossy. Moths emerge in August.

140 The Gothic

Family NOCTUIDAE *Naenia typica*

Food-plants Blackthorn, hawthorn, apple, sallow and a wide range of low-growing plants such as dock, willowherb and goosegrass
Season August–May
Distribution Widespread throughout the British Isles, and often common
Notes The caterpillars are frequently found in gardens where they will eat both weeds and cultivated plants. They are gregarious when young, and at this stage often feed together on the leaves of bushes, eating only the surface. Feeding takes place at night, the caterpillars concealing themselves at the base of the food-plant or under leaf litter on the ground by day. In October they go into hibernation, sheltering under stones or leaves. They become active again in the spring and resume feeding. Pupation takes place just below the surface of the soil within a thin cocoon of silk and earth. The glossy pupa is a dark reddish brown with the cremaster black. Moths emerge in June.

141 The Red Chestnut

Family NOCTUIDAE *Cerastis rubricosa*

Food-plants Low-growing plants such as dock, chickweed, groundsel and dandelion
Season April–June
Distribution Widespread throughout the British Isles

Notes The caterpillars feed at night, but remain on the food-plant by day, stretched out on the stems or leaves. When full fed they pupate within fragile earthen cocoons below the surface of the soil. The pupa is reddish brown and glossy. Moths emerge in March of the following year.

142 The Beautiful Yellow Underwing

Family NOCTUIDAE *Anarta myrtilli*

Food-plant Heather
Season July–October
Distribution Heaths and moorlands throughout the British Isles
Notes The caterpillars feed by day as well as at night, and often rest on the upper twigs of the food-plant. The broken green and white patterning of these caterpillars merges so successfully with the heather that they are very difficult to detect. When full grown they pupate within cocoons of silk covered with particles of earth, either on or below the surface of the soil and often attached to a root or stone. The glossy pupa is dark olive green with the abdomen dark reddish brown. Moths usually emerge in May or June.

143 The Nutmeg

Family NOCTUIDAE *Discestra trifolii*

Food-plants Goosefoot, Good King Henry, knotgrass and other plants
Season June or July–September
Distribution Widespread in England and Wales, becoming less common further north. It is scarce in Scotland and rare in Ireland
Notes The caterpillars feed at night, hiding by day on the ground under the food-plant. There are green and brown forms of this caterpillar, but both have

the characteristic white edged pink stripe along the line of the spiracles on each side. Pupation takes place just below the surface of the soil in a cocoon of earth and silk. The pupa is yellowish brown and smooth. Moths from autumn pupae emerge in May and June of the following year, but caterpillars hatching in early summer may produce moths in July and August.

144 The Cabbage Moth

Family NOCTUIDAE *Mamestra brassicae*

Food-plants Almost any herbage, but especially cabbage and related plants
Season July–September
Distribution Widespread, and often common throughout the British Isles
Notes Caterpillars of this moth are sometimes serious pests of cabbage crops and also cause damage to garden plants. The young caterpillars are greyish brown at first but change to a pale green colour as they develop. After the final moult they are often brown, although some remain green or grey-green until full grown. They feed at night, hiding by day on the ground beneath the food-plant. However, those on cabbage bore into the heart and feed inside where they remain concealed. Pupation takes place beneath the ground without the formation of a cocoon. The glossy pupa is pale reddish brown. Moths emerge in June of the following year, although sometimes caterpillars feed up quickly to produce moths in the autumn.

145 The Dot Moth

Family NOCTUIDAE *Melanchra persicariae*

Food-plants A wide range of low-growing plants and trees and shrubs

Season July–September
Distribution Widespread in England, Wales and Ireland, but not common in northern England, and of doubtful occurrence in Scotland
Notes These caterpillars are often common in gardens where they feed on both wild and cultivated plants. They feed at night, hiding on the undersides of leaves by day. The colour of these caterpillars varies from dark green to pale brown. Pupation takes place within a cocoon of earth and silk below the surface of the ground. The pupa is dark reddish brown and glossy. Moths usually emerge in July of the following year.

146 The Bright-line Brown-eye

Family NOCTUIDAE *Lacanobia oleracea*

Food-plants Many low-growing plants such as goosefoot, stinging nettle and dock. Also tomato
Season July–September
Distribution Widespread throughout the British Isles
Notes Although usually found on wild plants, the caterpillars are sometimes pests of tomato plants in glasshouses, and for this reason the species is known to growers as the Tomato Moth. The caterpillars feed at night and also to some extent by day, although at this time they usually hide on the ground beneath the food-plant. In the afternoons they sometimes sun themselves on the upper parts of the food-plant. They are extremely variable in colour, ranging from green to brown, with many intermediate forms including some striking pink varieties. The glossy black pupa is formed within a subterranean cocoon of earth and silk. Moths usually emerge in June of the following year.

147 The Broom Moth

Family NOCTUIDAE　　　　　　　　*Ceramica pisi*

Food-plants　Broom, bracken, bramble, sallow and many other plants
Season　August–September
Distribution　Widespread throughout the British Isles, and often common
Notes　The caterpillars feed mainly at night. By day they sometimes hide amongst the foliage, but at other times may be seen stretched out in conspicuous positions on ferns and other plants. There are a number of colour forms ranging from green to purplish black, but the characteristic yellow stripes are always present. The dark reddish brown pupa is formed within a subterranean cocoon made of earth. Moths emerge in June of the following year.

148 The Antler Moth

Family NOCTUIDAE　　　　　　　　*Cerapteryx graminis*

Food-plants　Grasses, such as purple moor-grass, mat-grass and Yorkshire fog
Season　March–June
Distribution　Throughout the British Isles, although less common in the south of England
Notes　These caterpillars prefer the harder and smoother species of grass. They feed on the leaves at night, hiding amongst the roots by day. In favourable seasons they can occur in huge numbers and sometimes devastate large areas of grassland on moors and hillsides. When full fed they pupate in earthen cocoons below the soil or amongst grass roots. The pupa is blackish brown and glossy. Moths emerge in August.

149 The Pine Beauty

Family NOCTUIDAE *Panolis flammea*

Food-plants Pine, particularly Scots pine
Season May–July
Distribution Throughout England, Wales and Scotland wherever pines are grown. It is scarce in Ireland
Notes The young caterpillars are pale yellow or green without markings, but as they grow the characteristic white stripes appear. When the striped pattern is fully developed they are very difficult to detect as they merge so successfully with the tufts of pine needles. The caterpillars eat the needles, starting at the tip and working down to the base. When full grown they pupate within frail silken cocoons under fallen pine needles or in the crevices of the bark. The pupa is reddish brown with the cremaster black. Moths emerge in March and April of the following year.

150 The Blossom Underwing

Family NOCTUIDAE *Orthosia miniosa*

Food-plants Mainly oak, but sometimes hawthorn, birch, bramble and low-growing herbaceous plants
Season May–June
Distribution Most common in southern England, but also occurs in Wales and as far north as Cumbria. It does not appear to occur in Scotland and is rare in Ireland
Notes The young caterpillars are gregarious, living at first under a common web and remaining in companies until well grown before dispersing. Webs are usually found on oak scrub rather than on trees. The caterpillars feed by day as well as at night and may

become full grown in a month. Pupation takes place beneath the surface of the soil or under leaf litter in an earthen cocoon. The pupa is a pale reddish brown in colour. Moths emerge in March of the following year.

151 The Common Quaker

Family NOCTUIDAE *Orthosia stabilis*

Food-plants Oak, birch, elm, beech, sallow and other trees
Season May–July
Distribution Widespread, and often common throughout the British Isles
Notes The caterpillars feed by day as well as at night, and may be found crawling on the tree-trunk. When full grown they pupate in the soil, often at the base of the tree. The pupa is reddish brown or dark purplish brown and glossy. Moths emerge in March of the following year.

152 The Clouded Drab

Family NOCTUIDAE *Orthosia incerta*

Food-plants Hawthorn, blackthorn, oak, sallow, apple and other trees
Season May–July
Distribution Widespread throughout the British Isles
Notes These caterpillars are sometimes pests in orchards where they not only feed on the foliage but also damage the fruits of the apple. They feed mainly at night but remain on the foliage by day. When full grown they pupate within earthen cocoons beneath the surface of the soil. The pupa is reddish brown and

glossy. Moths emerge in February or March of the following year.

153 The Hebrew Character

Family NOCTUIDAE *Orthosia gothica*

Food-plants Low-growing plants such as dock and dandelion, and also many deciduous trees and shrubs
Season April–June
Distribution Widespread, and common throughout the British Isles
Notes The young caterpillar is pale green with a black head, becoming grey-green with a yellow head when full grown. It feeds at night but remains on the food-plant by day. When full fed it buries itself under the soil and makes an earthen cocoon within which it pupates. The pupa is dull reddish brown with lighter wing-cases. Moths emerge in March or April of the following year.

154 The Clay

Family NOCTUIDAE *Mythimna ferrago*

Food-plants Chickweed, dandelion, plantain and grasses
Season September–April or May
Distribution Widespread in woodlands throughout the British Isles
Notes The caterpillars are most frequently found, after hibernation, feeding on grasses at the edge of woodland rides. They hide at the roots of the plants by day, climbing up to feed as soon as it is dark. When full grown they pupate within earthen cocoons below the surface of the soil. The pupa is a bright reddish brown and glossy. Moths emerge in June or July.

155 The Common Wainscot

Family NOCTUIDAE *Mythimna pallens*

Food-plants Grasses such as cock's-foot, annual meadow grass and couch grass
Season August–May
Distribution Widespread throughout the British Isles and often common
Notes The caterpillars feed externally on grass blades at night, hiding at the roots by day. When full grown in the spring they pupate between grass blades. The pupa is yellowish brown in colour. Moths emerge in June.

156 The Mullein Moth

Family NOCTUIDAE *Cucullia verbasci*

Food-plants Mullein, figwort and, in gardens, buddleia
Season June–July
Distribution Widespread in England and Wales, especially in the south. Not recorded from Scotland, and very local in Ireland
Notes The conspicuous caterpillars feed by day usually eating the leaves in preference to the flowers. It seems likely that these caterpillars are distasteful to birds, but if attacked they will quickly drop to the ground. When full grown they pupate within large cocoons constructed of layers of earth and silk at some depth below the surface of the soil. The pupa is reddish brown with the wing-cases greenish brown. Moths sometimes emerge in the following April or May, but others pass a second winter in the pupa, and there are records of moths emerging after five years.

157 The Sword Grass

Family NOCTUIDAE *Xylena exsoleta*

Food-plants A wide range of low-growing plants such as dock, groundsel, restharrow and thistle; also trees such as hazel, poplar and lime
Season April or May–June
Distribution Widespread throughout the British Isles, but more common in Scotland and northern England
Notes The caterpillars feed by day as well as at night and may be found feeding in bright sunlight. Although they usually feed on low-growing plants they may also eat young leaves of trees. When full grown the caterpillars pupate within hard earthen cocoons below the surface of the ground. The pupa is bright reddish brown and glossy. Moths emerge in September or October.

158 The Merveille du Jour

Family NOCTUIDAE *Dichonia aprilina*

Food-plant Oak
Season March–June
Distribution Widespread throughout the British Isles in localities where oak trees are plentiful
Notes The young caterpillar bores into an unopened bud, eating out the interior and sheltering within. Later it feeds on the opening buds, hiding by day in crevices in the bark. The colour and patterning of this caterpillar makes it very difficult to detect when at rest on a tree-trunk. When full fed it pupates within an earthen cocoon below the surface of the soil at the foot of the tree. The pupa is dull brown with the cremaster blackish brown. Moths usually emerge in September.

159 The Satellite

Family Noctuidae *Eupsilia transversa*

Food-plants Oak, beech, elm and other trees, and sometimes low-growing herbaceous plants
Season May–June
Distribution Widespread throughout the British Isles
Notes The young caterpillar draws leaves of its food-plant together with silk to form a shelter. These caterpillars are notoriously cannibalistic and so must be isolated when reared in captivity. When full grown they pupate below the surface of the soil within frail cocoons. The pupa is yellowish brown in colour. Moths usually emerge in September.

160 The Chestnut

Family Noctuidae *Conistra vaccinii*

Food-plants Oak, elm, sallow and low-growing herbaceous plants
Season May–June
Distribution Widespread in woodlands throughout the British Isles
Notes When young the caterpillars are semi-translucent and pale pink in colour with few markings. They feed at night, hiding amongst the foliage by day. When on the foliage of trees they conceal themselves between touching leaves or in a curled leaf, but when on low-growing plants they simply hide on the undersides of the leaves. When full grown they pupate below the surface of the soil at the roots of the food-plant within thick silken cocoons. The pupa is bright reddish brown or orange and glossy. Moths emerge in October or November.

161 The Poplar Grey

Family NOCTUIDAE *Acronicta megacephala*

Food-plant Poplar
Season July–September
Distribution Widespread throughout the British Isles, but most common in southern England
Notes The caterpillar feeds by day on the foliage of various types of poplar. When at rest on the upper surface of a leaf it assumes a characteristic 'hairpin' pose with the head turned to one side so that it touches the body. Pupation takes place within a strong cocoon of silk mixed with fragments of wood, concealed in a crevice of the bark. The pupa is reddish brown and glossy. Moths usually emerge in the following May or June, but sometimes remain in the pupa stage for a second winter.

162 The Sycamore

Family NOCTUIDAE *Acronicta aceris*

Food-plants Sycamore, maple, oak and sometimes plum and chestnut
Season August–September
Distribution Common throughout southern and eastern England, but of doubtful occurrence elsewhere in the British Isles
Notes At first the young caterpillar eats only the under-surface of the leaf, but later feeds in the usual way, consuming whole leaves. If disturbed it curls into a tightly coiled ring. The full-fed caterpillars may be found wandering on tree-trunks or palings and are often found crawling across paths. Pupation takes place within a double cocoon spun up on or under the bark. The outer layer of the cocoon is formed of loosely woven brownish silk mixed with body hairs and wood fragments while the inner layer

is of more densely woven silk. The pupa is brown with a darker line along the back. Moths emerge in June of the following year.

163 The Miller

Family NOCTUIDAE *Acronicta leporina*

Food-plants Birch, alder and sometimes oak or poplar
Season July–September
Distribution Widespread, but local throughout the British Isles
Notes The remarkable caterpillars of this moth may be found resting on the undersides of leaves. Southern specimens are usually pale green with white hairs, whilst those from the north are often yellowish with yellow hairs. Before pupation the full-fed caterpillar bores into dead wood or bark to form a shelter which it covers with silk interwoven with body hairs. The pupa is dark brown in colour with the wing veins showing dark green. Moths sometimes emerge in the following May, but often remain in the pupa stage for as long as three years.

164 The Alder Moth

Family NOCTUIDAE *Acronicta alni*

Food-plants A wide range of trees and bushes, including alder, hawthorn, blackthorn, birch, oak and sallow
Season July–September
Distribution Widespread, but local in England and Wales. In southern Ireland it is not uncommon in some areas
Notes In its early stages, the brown and white colour pattern of the caterpillar makes it resemble a bird-dropping but, after the final moult, it takes on its characteristic black and yellow coloration with

conspicuous paddle-shaped hairs. When full grown, the caterpillar excavates a shelter in rotten wood or soft bark within which it pupates. The pupa is dark reddish brown in colour. Moths emerge in June or July of the following year.

165 The Grey Dagger

Family NOCTUIDAE *Acronicta psi*

Food-plants Various trees and shrubs such as lime, hawthorn, blackthorn, and sometimes garden rose
Season August–September
Distribution Widespread throughout the British Isles, and often common
Notes The caterpillars are often found wandering on the trunks of trees by day. When full grown they pupate within compact whitish silken cocoons. These are usually hidden in crevices of the bark or in holes in rotten wood. The pupa is a deep reddish brown and glossy. Moths emerge in June of the following year.

166 The Marbled Beauty

Family NOCTUIDAE *Cryphia domestica*

Food-plant Lichen
Season August–May
Distribution Widespread in England, Wales, southern Scotland and Ireland
Notes The caterpillar constructs a cocoon-like shelter of silk covered with fragments of lichen, moss and other debris within which it hides for most of the day. Feeding usually takes place in the early morning. When full grown it spins a fresh shelter amongst

the lichen within which it pupates. The pupa is a light shining brown. Moths emerge in July.

167 The Copper Underwing

Family NOCTUIDAE *Amphipyra pyramidea*

Food-plants   Oak, birch, sallow and other trees and shrubs
Season   April–June
Distribution   Local in England and Wales, becoming more scarce in the north. It is not recorded from Scotland but is widespread in Ireland
Notes   The caterpillars feed at night but remain on the food-plant by day. When at rest they raise the head and thorax upright and spread out the legs. Pupation takes place within an earthen cocoon on the ground. The pupa is brown and glossy. Moths usually emerge in August.

168 The Old Lady

Family NOCTUIDAE *Mormo maura*

Food-plants   In autumn: chickweed, dock and many other low-growing plants. In spring: hawthorn, birch, sallow and fruit trees
Season   September–May
Distribution   Widespread throughout the British Isles, but most common in southern England
Notes   After feeding on the foliage of low-growing plants, the caterpillars hibernate while quite small. In spring they feed on the newly developed foliage of various trees and bushes. They feed at night, hiding on the ground by day. Pupation takes place within a large but fragile cocoon of earth and silk just below the surface of the ground. The large pupa is reddish brown with the surface powdered bluish white. Moths emerge in July.

169 The Angle Shades

Family NOCTUIDAE *Phlogophora meticulosa*

Food-plants A wide range of low-growing plants such as dock, groundsel and bracken
Season At any time of year
Distribution Widespread throughout the British Isles, and often common
Notes Although this moth is probably double brooded, caterpillars develop at different rates so that they may be found in any month of the year. They are commonly found in gardens where they feed on both weeds and cultivated plants and sometimes cause damage to plants in glasshouses. They feed at night, usually hiding beneath the plants or on the undersides of leaves by day. Pupation takes place within a rather fragile cocoon on the ground. The pupa is dark purplish brown and glossy. Moths emerge at all times of the year.

170 The Dun Bar

Family NOCTUIDAE *Cosmia trapezina*

Food-plants A wide range of deciduous trees such as oak, birch, elm and sallow
Season April–June
Distribution Widespread throughout the British Isles. It becomes less frequent in northern Scotland, but elsewhere is often common
Notes This caterpillar is a notorious cannibal and should be isolated immediately in captivity. It is probable that other caterpillars form a major part of its diet in the wild, particularly those of the Winter Moth. Pupation takes place within a loose silken cocoon on or just below the surface of the ground. The pupa is a dark reddish brown covered with a bluish bloom. Moths usually emerge in July.

171 The Dark Arches

Family NOCTUIDAE *Apamea monoglypha*

Food-plants Annual meadow-grass and other grasses
Season August–September
Distribution Widespread, and often common throughout the British Isles
Notes The caterpillar eats the stems of grasses just above the roots. It feeds at night, hiding by day in a small chamber among the grass roots. When full grown it buries itself just below the surface of the ground before pupating. The pupa is reddish brown in colour. Moths usually emerge in June or July of the following year.

172 The Common Rustic

Family NOCTUIDAE *Mesapamea secalis*

Food-plants Various grasses such as cock's-foot and annual meadow-grass
Season September–April or May
Distribution Common throughout the British Isles
Notes The caterpillars feed within the stems of grasses, and are sometimes pests of wheat. The presence of a caterpillar may be detected by a small round hole near the tip of a grass shoot through which the frass pellets are extruded. When full fed it burrows beneath the soil before pupating. The pupa is reddish brown in colour. Moths usually emerge in July.

173 The Rosy Rustic

Family NOCTUIDAE *Hydraecia micacea*

Food-plants Dock, plantain, potato, wheat and many other plants

Season May–August
Distribution Widespread throughout the British Isles
Notes The caterpillars are particularly fond of dock, feeding in the stems and boring down into the thick roots. They are sometimes pests of potato and tomato, hollowing out the stems. The pale yellow brown pupa is formed in the earth close to the roots of the food-plant. Moths emerge in September.

174 The Green Silver Lines

Family NOCTUIDAE *Pseudoips fagana*

Food-plants Oak, birch, beech and hazel
Season August–September
Distribution Widespread throughout the British Isles
Notes The caterpillars feed on the foliage both by day and at night. When full fed they pupate within papery, boat-shaped cocoons. Cocoons may be spun up in curled leaves, in crevices of the bark or amongst leaf litter on the ground. The pupa is purplish above, merging into orange-brown beneath; the wing-cases are yellowish brown. Moths usually emerge in June of the following year.

175 The Burnished Brass

Family NOCTUIDAE *Diachrysia chrysitis*

Food-plants Stinging nettle, dead-nettle and burdock
Season September–April and sometimes June–July or August
Distribution Widespread throughout the British Isles
Notes The caterpillars normally hatch in autumn, hibernating while quite small amongst dead leaves

on the ground and completing their development in the following spring. In favourable seasons, however, a second brood of caterpillars may be found feeding in the summer. Pupation takes place within a soft, whitish silken cocoon spun amongst leaves of the food-plant. The pupa is black with a long tongue case. Moths from overwintered caterpillars emerge in June or July, and those of the second brood in September.

176 The Golden Plusia

Family Noctuidae *Polychrysia moneta*

Food-plants Monkshood, delphinium and globe-flower
Season May–June and sometimes July–August
Distribution Widespread in England, Wales and Scotland. In Ireland only recorded from Co. Dublin
Notes The young caterpillars are greyish black with white dots, becoming dark bluish green dotted with black. At this stage they shelter amongst spun-together flower buds or in a spun leaf. Later, when the caterpillars attain their final yellowish green colour, they feed openly on the foliage or blossoms of the food-plant. Pupation takes place within an elongated dome-shaped cocoon of yellowish or white silk spun on the underside of a leaf. The pupa has the front surface greenish and the back blackish or reddish brown. Moths usually emerge in June, but sometimes a second brood emerges in August or September.

177 The Silver Y Moth

Family Noctuidae *Autographa gamma*

Food-plants A wide range of low-growing herbaceous plants
Season May–September

Distribution This migrant moth occurs throughout the British Isles

Notes The caterpillars of this moth are sometimes pests of clover, peas and other crops. Eggs are laid by immigrant moths in spring and early summer, and these produce caterpillars which feed through the summer to produce moths in the autumn. Further immigrants reinforce the population in autumn, but it is doubtful whether the moths survive the winter. Pupation takes place within a translucent white cocoon spun under a leaf or amongst debris. The pupa is black or very dark brown with a long and conspicuous tongue case.

178 The Red Underwing

Family Noctuidae *Catocala nupta*

Food-plants Willow and poplar
Season April–July
Distribution Occurs in south and east England as far north as Derbyshire. It is recorded from North Wales, but does not occur in Scotland or Ireland
Notes The caterpillars feed at night, hiding by day under bark or in crevices of the tree-trunk. The colour and patterning of the caterpillar makes it almost impossible to detect when at rest on the bark. When full grown it pupates within a tough brown cocoon between dead leaves or under loose bark. The pupa is purplish brown, powdered with a thick bluish white bloom. Moths emerge in August and September.

179 The Mother Shipton

Family Noctuidae *Callistege mi*

Food-plants Clover and field melilot
Season July–September
Distribution Widespread throughout England, Wales, Ireland and southern Scotland

Notes The slender caterpillar walks with a looping motion like those of the family Geometridae. When disturbed it rolls into a ball and remains motionless. Pupation takes place within a dull brown silken cocoon spun on a blade of grass which is twisted around it. The pupa is dark reddish brown covered with a whitish powder. Moths emerge in May of the following year.

180 The Herald Moth

Family NOCTUIDAE *Scoliopteryx libatrix*

Food-plants Sallow and willow
Season June–August
Distribution Widespread throughout the British Isles
Notes The caterpillars feed mainly at night, but remain on the leaves of the food-plant by day. They prefer the upper leaves and may be detected by signs of feeding, although the long green caterpillars merge very successfully with their surroundings. Pupation takes place within a frail white silken cocoon spun amongst the leaves of the food-plant. The pupa is dull black or sometimes brown. Moths emerge in August and September.

181 The Snout

Family NOCTUIDAE *Hypena proboscidalis*

Food-plant Stinging nettle
Season May–June
Distribution Throughout the British Isles where nettles grow in quantity
Notes The caterpillar lives in a shelter formed by spinning together the edges of a leaf. When full grown it pupates within a whitish silken web amongst leaves which are spun together. The dark reddish brown pupa has long wing-cases. Moths emerge in June or

July and there is sometimes a second generation which emerges in autumn.

182 The Common Fan-foot

Family NOCTUIDAE *Polypogon strigilata*

Food-plants Dead leaves of oak and birch
Season August–April or May
Distribution Widespread in southern England, extending into the Midlands. In Ireland it is very rare
Notes The caterpillars are almost full grown before hibernation, but resume feeding for a short time in the spring. Although they usually feed on dead leaves they have been known to eat birch catkins in spring. Pupation takes place within a very fragile cocoon of silk mixed with debris on the ground or in a crevice of the bark. The pupa is reddish brown in colour. Moths emerge in May or June.

183 The Common Swift

Family HEPIALIDAE *Hepialus lupulinus*

Food-plants Grasses and most herbaceous plants
Season July–April
Distribution Widespread throughout the British Isles, and often common
Notes The caterpillars of this moth feed on the roots of a wide range of plants, and sometimes cause serious damage to field crops and to both flowers and vegetables in gardens. They feed below ground throughout the winter, although most damage is caused in early spring. The full-grown caterpillar pupates in an underground cocoon. The pale reddish brown pupa has short wing-cases and the abdominal segments bear tooth-like spines which enable it to work its way to the surface prior to the emergence of the moth. Moths usually emerge in June.

184 The Leopard Moth

Family COSSIDAE *Zeuzera pyrina*

Food-plants Maple, sycamore, ash, elm, oak and many others, including ornamental and fruit trees
Season At all times of year
Distribution Southern England, particularly the London area, and South Wales
Notes This wood-boring species is sometimes a pest and severely attacked trees may have to be destroyed. On hatching the young caterpillar feeds just below the bark where it remains throughout the winter. In spring it bores into the wood of the branch or stem making a round tunnel. The caterpillar feeds for two or three years before pupating in a cocoon of silk mixed with wood fragments, formed in the tunnel near to the bark. The pupa is reddish brown in colour and has spines along the abdominal segments which enable it to force its way out of the tunnel prior to the emergence of the moth. Moths usually emerge in June or July.

185 The Goat Moth

Family COSSIDAE *Cossus cossus*

Food-plants Various trees, especially elm, ash and willow, and sometimes fruit trees
Season At all times of year
Distribution Widespread throughout the British Isles, but rather uncommon in Ireland
Notes This wood-boring species can cause serious damage to a tree which is heavily infested. It occasionally attacks and kills fruit trees such as cherry, plum and apple. The strong goat-like smell of the caterpillar gives the moth its common name. It takes three or four years to complete its growth, boring tunnels in the solid wood of branches and tree-trunks. When full fed it either pupates in a tough cocoon of

silk and wood fragments at the entrance of its tunnel or leaves its burrow to pupate on the ground in a tough earthen cocoon lined with silk. The large pupa is reddish brown and glossy. Moths usually emerge in June.

186 The Forester

Family ZYGAENIDAE *Adscita statices*

Food-plant Sorrel
Season July–April
Distribution Widespread throughout the British Isles
Notes After hatching the young caterpillar bores into a leaf of the food-plant and feeds on the tissues between the upper and lower skins. Later it feeds on the underside of the leaf leaving the upper skin intact. After hibernating through the winter it completes its growth in the spring and pupates within a flimsy whitish cocoon spun low down amongst the herbage. The pupa is black and shining. Moths usually emerge in June.

187 The Six-spot Burnet

Family ZYGAENIDAE *Zygaena filipendulae*

Food-plant Bird's-foot trefoil
Season August–May
Distribution Widespread throughout the British Isles
Notes The caterpillars feed openly by day and are probably protected by the fact that they are distasteful to most predators. After hibernating through the winter they resume feeding in the spring and when full grown pupate in boat-shaped whitish or yellowish cocoons spun high up on the grass stems. The pupa is black and shining. Moths usually emerge in July.

188 The Five-spot Burnet

Family ZYGAENIDAE *Zygaena trifolii*

Food-plants Bird's-foot trefoil and marsh bird's-foot trefoil
Season July–May
Distribution Southern England, Wales and the Isle of Man
Notes There are two subspecies of this moth, one found on downland and the other in marshes. The former feeds on bird's-foot trefoil and when full grown spins its cocoon low down amongst the herbage. The marsh subspecies feeds on marsh bird's-foot trefoil and spins its cocoon well up on grass stems. In both cases the boat-shaped cocoon is whitish or yellowish and the glossy pupa is black or brown. Moths of the downland form emerge in May and June while those of the marsh form emerge in late July or early August.

189 The 'Bagworm'

Family PSYCHIDAE *Psyche casta*

Food-plants Grasses and probably lichens
Season July–May
Distribution Widespread throughout the British Isles with the exception of northern Scotland
Notes Although this moth does not have a generally accepted common name, its caterpillar is frequently found on walls and fences. The caterpillar lives within a portable silken case covered with longitudinally arranged fragments of grasses. It hibernates through the winter and resumes feeding in the spring before pupating within its case, which it attaches to a grass stem or other support with silk. Moths emerge in June or July. The females are wingless.

190 The Festoon

Family LIMACODIDAE *Apoda avellana*

Food-plant Oak
Season August–October
Distribution Oak woodlands in southern England
Notes The remarkably flattened, slug-like caterpillar of this moth becomes full grown in the autumn and spins a brownish silken cocoon which it attaches to a leaf of the food-plant. It remains within the cocoon throughout the winter and does not pupate until the spring. The pupa is pale brown in colour. Moths usually emerge in June.

191 The Currant Clearwing

Family SESIIDAE *Synanthedon salmachus*

Food-plants Blackcurrant and redcurrant
Season July–May
Distribution Widespread in England, but less common in Wales and southern Scotland. It is also recorded from various parts of Ireland
Notes Caterpillars of this moth are sometimes minor pests of currant bushes and occasionally attack gooseberry as well. The young caterpillar bores into the stem and feeds on the pith, gradually tunnelling towards the growing point of the shoot. It feeds through the winter and usually becomes full grown by April. Leaves of infested shoots wilt and fruit trusses fail to mature. When full fed, the caterpillar bores an escape hole to the side of the stem, leaving a thin outer skin through which the pupa can break just prior to the emergence of the moth. Pupation takes place within this tunnel. The pupa is bright reddish brown in colour. Moths emerge in June or July.

192 The Common Clothes Moth

Family TINEIDAE *Tineola bisselliella*

Food Wool, feathers, fur and sometimes stored foods
Season October–June
Distribution Widespread throughout the British Isles
Notes This is the once notorious clothes moth which did so much damage to woollen clothes, but today with increased use of synthetic fibres and improved hygiene is a less serious pest. The caterpillars prefer to feed on soiled wool and seldom do extensive damage to clean clothes. The small pale brown pupa is formed within a tough silken cocoon spun amongst the food. Moths usually emerge in the summer, but may occur at any time of year.

193 The Case-bearing Clothes Moth

Family TINEIDAE *Tinea pellionella*

Food Wool, feathers, fur and other materials of animal origin
Season August–May
Distribution Widespread throughout the British Isles
Notes Like the Common Clothes Moth, this species is now becoming much less common, but sometimes damages neglected woollen carpets and other fabrics. Out of doors it occurs in birds' nests where it feeds on feathers and other debris. The caterpillar lives within a portable silken case interwoven with fragments of the material on which it is feeding. When full grown it attaches the case to a suitable surface with silk and pupates inside. The pupa is brown in colour. Moths emerge in summer or autumn.

194 The Small Ermine Moth

Family YPONOMEUTIDAE *Yponomeuta padella*

Food-plants Hawthorn and blackthorn
Season August–June
Distribution Widespread throughout the British Isles
Notes Although the caterpillars hatch in late summer or autumn, they remain under the protection of the egg mass on a branch of the food-plant until the following spring. They then spin a communal web over the developing foliage and feed within, extending the web as the leaves are consumed. In cases of severe infestation, entire trees or hedges may be enveloped in silk and subsequently defoliated. The full-grown caterpillars pupate within the web, often spinning their cocoons together. The pupa is pale yellowish brown with the thorax, wing-cases and tip of the abdomen dark brown. Moths emerge in July and August. A number of closely related species with similar appearance and habits feed on a wide range of shrubs and trees, including fruit trees. They are also known as 'small ermine moths'.

195 The Diamond-back Moth

Family YPONOMEUTIDAE *Plutella xylostella*

Food-plants Cabbage and other related plants
Season June–July and August–September
Distribution A migrant species occurring throughout the British Isles
Notes This moth is double brooded in the British Isles, but is reinforced by migrants from the Continent. It can be a serious pest of cabbage crops, particularly along the east coast of England. The caterpillars live in a slight web and feed on the undersides of the leaves making characteristic holes. They

become full grown in about three weeks and spin small open network cocoons on the food-plant. The pupa is pale green in colour. Moths of the first brood emerge in August, while those of the second brood emerge in the following May.

196 The Brown House Moth

Family OECOPHORIDAE *Hofmannophila pseudospretella*

Food Seeds, dried plants, wool and other materials of animal origin
Season June–April
Distribution Widespread throughout the British Isles
Notes This is probably the most common moth pest in houses. The caterpillars usually feed amongst dust and debris behind skirting and between floor boards, and often damage carpets. They also live in old birds' nests and wasps' nests, and where these occur in lofts they are often a source of infestation. When full grown, the caterpillar pupates within a cocoon made of silk mixed with particles of food or debris. The pupa is pale brown in colour. Moths usually emerge in summer or autumn.

197 The Pea Moth

Family TORTRICIDAE *Cydia nigricana*

Food-plants Vetches and wild and cultivated pea
Season July–September
Distribution Widespread in the British Isles
Notes The caterpillars of this moth are a serious crop pest and are the familiar 'maggots' found in pea pods. On hatching they bore into the young pods and feed inside on the developing seeds. When full grown they leave the pods and descend into the soil where they pupate below the surface. The pupa is yellowish

brown in colour. Moths emerge in June or July of the following year.

198 The Codling Moth

Family TORTRICIDAE *Cydia pomonella*

Food-plants Apple, pear, quince and walnut
Season July–May
Distribution Widespread wherever apples are grown
Notes The caterpillar of this common pest moth is the well known 'maggot' of apples. On hatching, the young caterpillar bores into the fruit until it reaches the core where it feeds on the pips and surrounding flesh. After about a month, when it is full fed, the caterpillar leaves the apple and spins a cocoon under loose bark or in a suitable crevice where it spends the winter before pupating in the spring. The pupa is brown in colour. Moths usually emerge in late spring or summer, but sometimes a small second brood emerges in autumn.

199 The Carnation Tortrix

Family TORTRICIDAE *Cacoecimorpha pronubana*

Food-plants A wide range of low-growing plants, trees and shrubs, including strawberry, rose, privet, greenhouse carnations and many house plants
Season May–August and September–April
Distribution Widespread and sometimes common in southern England, extending into Wales
Notes In recent years, this moth has established itself in gardens and is sometimes a pest of cultivated flowers and vegetables, especially those grown under glass. The caterpillar spins the leaves of the food-plant together and feeds within the shelter formed. When full grown it pupates in a spun leaf. The pupa is dark brown with black wing-cases and thorax.

Moths from the summer caterpillars usually emerge in autumn, whilst those of the second generation emerge in the following spring.

200 The Green Oak Tortrix

Family TORTRICIDAE *Tortrix viridana*

Food-plants Usually oak, but also beech, sycamore, poplar, sallow and bilberry
Season April–June
Distribution Widespread throughout the British Isles
Notes On hatching from the overwintered eggs the young caterpillars enter the buds in which they feed. Later they eat the developed foliage, living within folded or rolled leaves. Sometimes the moth occurs in such abundance that whole trees are defoliated by the caterpillars. In such cases they sometimes drop to the ground and feed on the undergrowth. When full grown after about a month they pupate within spun leaves of the food-plant. The pupa is black or brown in colour. Moths emerge in June or July.

201 The Garden Pebble

Family PYRALIDAE *Evergestis forficalis*

Food-plants Cabbage, cauliflower, turnip, radish and related plants
Season June–July and September–April
Distribution Widespread throughout the British Isles
Notes Caterpillars of this moth are sometimes a pest of cabbages in gardens. There are two broods a year, and occasionally a small third generation is produced. They usually attack the hearts of cabbages and sometimes web the young leaves together. They feed only at night. When full grown the caterpillars enter the soil and spin tough silken cocoons under-

ground. The pupa is brown in colour. Caterpillars of the second brood remain in the cocoon until the spring before pupation and produce moths in May or June. Summer caterpillars pupate in July to produce moths in August or September.

202 The Small Magpie Moth

Family PYRALIDAE *Eurrhypara hortulata*

Food-plants Usually nettle, but also other plants such as horehound, woundwort and mint
Season August–April
Distribution Widespread throughout the British Isles with the exception of northern Scotland
Notes The caterpillars live and feed within rolled and spun leaves of nettle. When full grown in autumn they leave the food-plant to find a suitable pupation site and, in cases where large nettle beds grow nearby, they may enter houses at this stage. The caterpillars remain within their translucent white silken cocoons throughout the winter before pupating in the spring. The pupa is reddish brown in colour. Moths emerge in June or July.

203 The Warehouse Moth

Family PYRALIDAE *Ephestia elutella*

Food Chocolate, dried fruit, nuts, cereals and other stored foods
Season July–April
Distribution In houses, shops and warehouses throughout the British Isles
Notes This is one of the most serious pests of stored foods in Britain. The caterpillars feed within the food, webbing it together with fine threads of silk. They are often overlooked until they are full grown when

they leave the food to find suitable pupation sites in the packing material and cover the outside of the containers with a thin sheet of silk. The yellowish brown pupae are formed within whitish silken cocoons. Some caterpillars pupate in September to produce moths in autumn, but most overwinter before pupating in spring to produce moths in early summer.

Bibliography

Allen, P. B. M., *Larval Food-plants*. Watkins & Doncaster, 1949.

Dickson, R., *A Lepidopterist's Handbook*. Amateur Entomologist's Society, 1976.*

Ford, E. B., *Butterflies*. New Naturalist Series No. 1, Collins, 1945.

Ford, E. B., *Moths*. New Naturalist Series No. 30, Collins, 1955.

Ford, R. L. E., *Studying Insects*. Warne, 1973.

Ford, R. L. E., *The Observer's Book of Larger Moths*. Warne, 1974.

Howarth, T. G., *Colour Identification Guide to British Butterflies*. Warne, 1973.

Howarth, T. G., *South's British Butterflies*. Warne, 1973.

South, R., *Moths of the British Isles*. Warne, revised edition 1961.

Stokoe, W. J., *The Observer's Book of Butterflies*. Warne, revised edition 1969.

* The Amateur Entomologist's Society also publish a wide range of leaflets and booklets on all aspects of collecting and studying insects. A list is available from The AES Publications Agent, 129 Franciscan Road, Tooting, London SW17 8DZ.

Index

Figures in italics refer to plates

Abraxas grossulariata 82, Pl 11: *70*
Acherontia atropos 92, Pl 15: *89*
Acronicta aceris 131, Pl 27: *162*
Acronicta alni 132, Pl 27: *164*
Acronicta leporina 132, Pl 27: *163*
Acronicta megacephala 131, Pl 27: *161*
Acronicta psi 133, Pl 27: *165*
Adscita statices 143, Pl 31: *186*
Aglais urticae 54, Pl 4: *23*
Agrotis puta 114, Pl 21: *129*
Agrotis segetum 114, Pl 27: *128*
Alcis repandata 89, Pl 14: *83*
Alder Moth 132, Pl 27: *164*
Alsophila aescularia 75, Pl 9: *56*
Amphipyra pyramidae 134, Pl 28: *167*
Anarta myrtilli 121, Pl 23: *142*
Angle Shades 135, Pl 28: *169*
Antler Moth 124, Pl 25: *148*
 as pest 27
Anthocharis cardamines 47, Pl 2: *11*
Apamea monoglypha 136, Pl 28: *171*
Apanteles wasps 23
Apatura iris 53, Pl 4: *20*
Apeira syringaria 84, Pl 12: *73*
Aphantopus hyperantus 64, Pl 6: *38*
Apocheima pilosaria 86, Pl 12: *77*
Apoda avellana 145, Pl 31: *190*
Archiearis parthenias 74, Pl 9: *55*
Arctia caja 108, Pl 20: *117*
Arctia villica 109, Pl 20: *118*
ARCTIIDAE 106–13
Argynnis adippe 57, Pl 5: *28*
Argynnis aglaja 58, Pl 5: *29*
Argynnis paphia 59, Pl 5: *30*
Autographa gamma 138, Pl 29: *177*
Axylia putris 115, Pl 22: *130*

'Bagworm' 144, Pl 31: *189*
Beating tray 28
Beautiful Yellow Underwing 121, Pl 23: *142*
Biological control 28
Biston betularia 87, Pl 13: *79*
Black Arches 106, Pl 19: *113*
Blossom Underwing 125, Pl 25: *150*
Boloria euphrosyne 57, Pl 5: *27*
Boloria selene 56, Pl 5: *26*
Bordered White 90, Pl 14: *86*
Bright-line Brown-eye 123, Pl 24: *146*
Brimstone Butterfly 44, Pl 2: *7*
Brimstone Moth 83, Pl 12: *72*
Brindled Beauty 87, Pl 13: *78*
Broom Moth 126, Pl 25: *147*
Broad-bordered Yellow Underwing 116, Pl 22: *133*
Brown House Moth 148, Pl 32: *196*
Brown-tail Moth 104, Pl 19: *110*
 hairs of 19, 27
Buff Arches 73, Pl 9: *52*
Buff Ermine 110, Pl 20: *121*
Buff-tip 96, Pl 16: *96*
Bupalus piniaria 90, Pl 14: *86*
Burnet moths, defence 20
Burnished Brass 137, Pl 29: *175*

Cabbage Moth 122, Pl 24: *144*
Cacoecimorpha pronubana 149, Pl 32: *199*
Callimorpha dominula 112, Pl 21: *124*
Callistege mi 139, Pl 30: *179*
Callophrys rubi 48, Pl 3: *12*

153

Camouflage 20–22
Camptogramma bilineata 78, Pl 10: 62
Cannibalism 30, 130, 135
Carnation Tortrix 149, *Pl 32: 199*
Case-bearing Clothes Moth 146, *Pl 31: 193*
Caterpillar, structure 15–18
Catocala nupta 139, *Pl 29: 178*
Celastrina argiolus 51, *Pl 3: 18*
Ceramica pisi 126, *Pl 25: 147*
Cerapteryx graminis 124, *Pl 25: 148*
Cerastis rubricosa 120, *Pl 23: 141*
Cerura vinula 97, *Pl 17: 97*
Chalkhill Blue Butterfly 50, *Pl 3: 17*
Chestnut 130, *Pl 27: 160*
Chinese Character 71, *Pl 9: 50*
Chitin 11–12
Chloroclysta truncata 79, *Pl 11: 64*
Chloroclystis rectangulata 82, *Pl 11: 69*
Chorionin 14
Chrysalis 12
Cilix glaucata 71, *Pl 9: 50*
Cinnabar Moth 112, *Pl 21: 125*
  warning coloration 20
Clay 127, *Pl 26: 154*
Clothes moths 26, 146
Clostera pigra 101, *Pl 18: 105*
Clouded Buff 109, *Pl 20: 119*
Clouded Drab 126, *Pl 25: 152*
Clouded Yellow Butterfly 44, *Pl 2: 6*
Cocoon 13
Codling Moth, 26, 149, *Pl 32: 198*
Coenonympha pamphilus 63, *Pl 6: 37*
Colias croceus 44, *Pl 2: 6*
Colotois pennaria 86, *Pl 12: 76*
Collecting methods 28–31
Comma Butterfly 56, *Pl 5: 25*
Common Blue Butterfly 50, *Pl 3: 16*
Common Clothes Moth 146, *Pl 31: 192*
Common Emerald 76, *Pl 10: 59*
Common Fan-foot 141, *Pl 30: 182*

Common Footman 107, *Pl 20: 116*
Common Heath 90, *Pl 14: 85*
Common Lutestring 73, *Pl 9: 53*
Common Marbled Carpet 79, *Pl 11: 64*
Common Pug 81, *Pl 11: 68*
Common Quaker 126, *Pl 25: 151*
Common Rustic 136, *Pl 28: 172*
Common Swift 141, *Pl 30: 183*
  as pest 26
Common Wainscot 128, *Pl 26: 155*
Conistra vaccinii 130, *Pl 27: 160*
Conservation 24–5, 38
Copper Underwing 134, *Pl 28: 167*
Cosmia trapezina 135, *Pl 28: 170*
COSSIDAE 142
Cossus cossus 142, *Pl 30: 185*
Coxcomb Prominent 100, *Pl 18: 104*
Cream-spot Tiger 109, *Pl 20: 118*
Cremaster 13
Cryphia domestica 133, *Pl 28: 166*
Cucullia verbasci 128, *Pl 26: 156*
Cupido minimus 49, *Pl 3: 15*
Currant Clearwing 145, *Pl 31: 191*
Cutworm caterpillars 25–6
Cydia nigricana 148, *Pl 32: 197*
Cydia pomonella 26, 149, *Pl 32: 198*
Cylinder cages 34
Cynthia cardui 54, *Pl 4: 22*

Dark Arches 136, *Pl 28: 171*
Dark Green Fritillary 58, *Pl 5: 29*
Dark Spinach 69, *Pl 10: 63*
Dark Tussock 103, *Pl 18: 108*
Dasychira fascelina 103, *Pl 18: 108*
Dasychira pudibunda 103, *Pl 19: 109*
Death's Head Hawk-moth 92, *Pl 15: 89*
December Moth 65, *Pl 7: 39*
Defence 18–22

154

Deilephila elpenor 95, *Pl 16: 95*
Diachrysia chrysitis 137, *Pl 29: 175*
Diacrisia sannio 109, *Pl 20: 119*
Diamond-back Moth 147, *Pl 32: 195*
Diaphora mendica 111, *Pl 20: 122*
Diarsia brunnea 118, *Pl 23: 136*
Diarsia mendica 117, *Pl 22: 135*
Dichonia aprilina 129, *Pl 26: 158*
Diloba caeruleocephala 101, *Pl 18: 106*
Dingy Footman 107, *Pl 19: 115*
Dingy Skipper Butterfly 42, *Pl 1: 3*
Discestra trifolii 121, *Pl 24: 143*
Diseases 24
Dot Moth 122, *Pl 24: 145*
Drepana falcataria 71, *Pl 9: 49*
DREPANIDAE 71–2
Drinker Moth 68, *Pl 8: 45*
Dun Bar 135, *Pl 28: 170*
 as cannibal 30

Early Moth 91, *Pl 14: 87*
Early Thorn 84, *Pl 12: 74*
Ecdysis 12
Egg 14
Eilema griseola 107, *Pl 19: 115*
Eilema lurideola 107, *Pl 20: 116*
Elephant Hawk-moth 95, *Pl 16: 95*
 threat posture 19
Ematurga atomaria 90, *Pl 14: 85*
Emperor moth 70, *Pl 8: 47*
ENDROMIDAE 70–1
Endromis versicolora 70, *Pl 8: 48*
Ephestia elutella 151, *Pl 32: 203*
Erannis defoliaria 88, *Pl 13: 80*
Eriogaster lanestris 66, *Pl 7: 41*
Erynnis tages 42, *Pl 1: 3*
Euproctis chrysorrhoea 104, *Pl 19: 110*
Euproctis similis 105, *Pl 19: 111*
Eupsilia transversa 130, *Pl 26: 159*
Eurrhypara hortulata 151, *Pl 32: 202*

Euxoa nigricans 113, *Pl 21: 127*
Evergestis forficalis 150, *Pl 32: 201*
Eyed Hawk-moth 94, *Pl 15: 92*
Eyes 16

Feathered Thorn 86, *Pl 12: 76*
Festoon 145, *Pl 31: 190*
Figure of Eight Moth 101, *Pl 18: 106*
Five-spot Burnet 144, *Pl 31: 188*
Flame Moth 115, *Pl 22: 130*
Forester 143, *Pl 31: 186*
Fox Moth 68, *Pl 7: 44*
Frosted Green 74, *Pl 9: 54*
Fruit, damage by caterpillars 26

Garden Carpet 77, *Pl 10: 61*
Garden Dart 113, *Pl 21: 127*
Garden Pebble 150, *Pl 32: 201*
Garden Tiger 108, *Pl 20: 117*
Gastropacha quercifolia 69, *Pl 8: 46*
Gatekeeper Butterfly 62, *Pl 6: 35*
Geometra papilionaria 76, *Pl 10: 58*
GEOMETRIDAE 74–92
 camouflage 20
Goat Moth 142, *Pl 30: 185*
Golden Plusia 138, *Pl 29: 176*
Gonepteryx rhamni 44, *Pl 2: 7*
Gothic 120, *Pl 23: 140*
Grass Emerald 75, *Pl 9: 57*
Grass Wave 91, *Pl 14: 88*
Grayling Butterfly 61, *Pl 6: 34*
Green Hairstreak Butterfly 48, *Pl 3: 12*
Green Oak Tortrix 150, *Pl 32: 200*
Green Pug 82, *Pl 11: 69*
Green Silver Lines 137, *Pl 29: 174*
Green-veined White Butterfly 46, *Pl 2: 10*
Grey Dagger 133, *Pl 27: 165*
Grey Pine Carpet 80, *Pl 11: 65*
Grizzled Skipper Butterfly 42, *Pl 1: 4*

Habrosyne pyritoides 73, *Pl 9: 52*
Hairs 19

Harpyia bifida 98, *Pl 17: 99*
Harpyia furcula 97, *Pl 17: 98*
Heath Rustic 119, *Pl 23: 139*
Hebrew Character 127, *Pl 26: 153*
Hemithea aestivaria 76, *Pl 10: 59*
HEPIALIDAE 141
Hepialus lupulinus 141, *Pl 30: 183*
Herald Moth 140, *Pl 30: 180*
HESPERIIDAE 41–3
High Brown Fritillary 57, *Pl 5: 28*
Hipparchia semele 61, *Pl 6: 34*
Hofmannophila pseudospretella 148, *Pl 32: 196*
Holly Blue Butterfly 51, *Pl 3: 18*
House moths 26
Humming-bird Hawk-moth 95, *Pl 16: 94*
Hydraecia micacea 136, *Pl 29: 173*
Hypena proboscidalis 140, *Pl 30: 181*

Identification 31
Inachis io 55, *Pl 4: 24*
Ingrailed Clay 117, *Pl 22: 135*
Instar 12
Iron Prominent 99, *Pl 17: 101*

Jodis lactearia 77, *Pl 10: 60*

Kentish Glory 70, *Pl 8: 48*

Lacanobia oleracea 123, *Pl 24: 146*
Lackey Moth 66, *Pl 7: 42*
Ladoga camilla 52, *Pl 4: 19*
Laothoe populi 94, *Pl 15: 93*
Lappet Moth 69, *Pl 8: 46*
Large Emerald 76, *Pl 10: 58*
Large Skipper Butterfly 41, *Pl 1: 2*
Large White Butterfly 45, *Pl 2: 8*
  as pest 26
  parasites of 23
Large Yellow Underwing 115, *Pl 22: 131*

Lasiocampa quercus 67, *Pl 7: 43*
LASIOCAMPIDAE 65–9
Lasiommata megera 60, *Pl 6: 32*
Legs 17–18
Leopard Moth 142, *Pl 30: 184*
Lesser Satin Moth 73, *Pl 9: 53*
Lesser Swallow Prominent 99, *Pl 18: 102*
Lesser Yellow Underwing 116, *Pl 22: 132*
Leucoma salicis 105, *Pl 19: 112*
Life cycle 11–14
Lilac Beauty 84, *Pl 12: 73*
LIMACODIDAE 145
Lime Hawk-moth 93, *Pl 15: 91*
Little Emerald 77, *Pl 10: 60*
Lobster Moth 98, *Pl 17: 100*
  defence 19–20
Looper caterpillars 18, 20, 21
Lycaena phlaeas 49, *Pl 3: 14*
LYCAENIDAE 48–52
Lycia hirtaria 87, *Pl 13: 78*
Lycophotia porphyrea 117, *Pl 22: 134*
Lymantria monacha 106, *Pl 19: 113*
LYMANTRIIDAE 102–6
Lysandra coridon 50, *Pl 3: 17*

Macroglossum stellatarum 95, *Pl 16: 94*
Macrothylacia rubi 68, *Pl 7: 44*
Magpie Moth 82, *Pl 11: 70*
  as pest 26
Malacosoma neustria 66, *Pl 7: 42*
Mamestra brassicae 122, *Pl 24: 144*
Maniola jurtina 62, *Pl 6: 36*
Marbled Beauty 133, *Pl 28: 166*
Marbled White Butterfly 61, *Pl 6: 33*
March Moth 75, *Pl 9: 56*
Meadow Brown Butterfly 62, *Pl 6: 36*
Melanargia galathea 61, *Pl 6: 33*
Melanchra persicariae 122, *Pl 24: 145*
Menophra abruptaria 88, *Pl 13: 81*
Merveille du Jour 129, *Pl 26: 158*

156

Mesapamea secalis 136, *Pl 28: 172*
Micropyle 14
Migration 14
Miller 132, *Pl 27: 163*
Mimas tiliae 93, *Pl 15: 91*
Mormo maura 134, *Pl 28: 168*
Mother Shipton 139, *Pl 30: 179*
Mottled Beauty 89, *Pl 14: 83*
Mottled Umber 88, *Pl 13: 80*
Moulting 12
Mouthparts 15
Mullein Moth 128, *Pl 26: 156*
Muslin Footman 106, *Pl 19: 114*
Muslin Moth 111, *Pl 20: 122*
Mythimna ferrago 127, *Pl 26: 154*
Mythimna pallens 128, *Pl 26: 155*

Naenia typica 120, *Pl 23: 140*
Noctua comes 116, *Pl 22: 132*
Noctua fimbriata 116, *Pl 22: 133*
Noctua pronuba 115, *Pl 22: 131*
NOCTUIDAE 113–41
Nola cucullatella 113, *Pl 21: 126*
NOLIDAE 113
Notodonta dromedarius 99, *Pl 17: 101*
NOTODONTIDAE 96–102
Nudaria mundana 106, *Pl 19: 114*
Nutmeg 121, *Pl 24: 143*
NYMPHALIDAE 52–9

Oak Eggar 67, *Pl 7: 43*
Ochlodes venata 41, *Pl 1: 2*
Ochropacha duplaris 73, *Pl 9: 53*
OECOPHORIDAE 148
Old Lady 134, *Pl 28: 168*
Operophtera brumata 80, *Pl 11: 66*
Opisthograptis luteolata 83, *Pl 12: 72*
Orange Underwing 74, *Pl 9: 55*
Orange-tip Butterfly 47, *Pl 2: 11*
Orgyia antiqua 102, *Pl 18: 107*
Orthosia gothica 127, *Pl 26: 153*
Orthosia incerta 126, *Pl 25: 152*
Orthosia miniosa 125, *Pl 25: 150*
Orthosia stabilis 126, *Pl 25: 151*
Osmeteria 19
Ourapteryx sambucaria 85, *Pl 12: 75*

Painted Lady Butterfly 54, *Pl 4: 22*
Pale Brindled Beauty 86, *Pl 12: 77*
Pale Oak Beauty 90, *Pl 14: 84*
Pale Oak Eggar 65, *Pl 7: 40*
Pale Tussock 103, *Pl 19: 109*
Panolis flammea 125, *Pl 25: 149*
Papilio machaon 43, *Pl 1: 5*
PAPILIONIDAE 43
Pararge aegeria 59, *Pl 6: 31*
Parasites of caterpillars 22–3
Pea Moth 148, *Pl 32: 197*
Peach Blossom 72, *Pl 9: 51*
Peacock Butterfly 55, *Pl 4: 24*
Pearl-bordered Fritillary 57, *Pl 5: 27*
Pebble Hook-tip 71, *Pl 9: 49*
Pelurga comitata 79, *Pl 10: 63*
Peppered Moth 87, *Pl 13: 79*
Perconia strigillaria 91, *Pl 14: 88*
Peribatodes rhomboidaria 89, *Pl 14: 82*
Perizoma didymata 81, *Pl 11: 67*
Pest species 25–8
Phalera bucephala 96, *Pl 16: 96*
Pheosia gnoma 99, *Pl 18: 102*
Pheosia tremula 100, *Pl 18: 103*
Philudoria potatoria 68, *Pl 8: 45*
Phlogophora meticulosa 135, *Pl 28: 169*
Phragmatobia fuliginosa 111, *Pl 21: 123*
PIERIDAE 44–7
Pieris brassicae 45, *Pl 2: 8*
Pieris napi 46, *Pl 2: 10*
Pieris rapae 45, *Pl 2: 9*
Pine Beauty 125, *Pl 25: 149*
Plutella xylostella 147, *Pl 32: 195*
Poecilocampa populi 65, *Pl 7: 39*
Polychrysia moneta 138, *Pl 29: 176*
Polygonia c-album 56, *Pl 5: 25*
Polyommatus icarus 50, *Pl 3: 16*
Polyploca ridens 74, *Pl 9: 54*
Polypogon strigilata 141, *Pl 30: 182*
Poplar Grey 131, *Pl 27: 161*
Poplar Hawk-moth 94, *Pl 15: 93*

Poplar Kitten 98, *Pl 17: 99*
Predation 22
Preservation 31–2
Privet Hawk-moth 93, *Pl 15: 90*
Prolegs 17–18
Pseudoips fagana 137, *Pl 29: 174*
Pseudoterpna pruinata 75, *Pl 9: 57*
Psyche casta 144, *Pl 31: 189*
PSYCHIDAE 20, 144
Ptilodon capucina 100, *Pl 18: 104*
Pupa 12–13
Pupation 36
Purple Clay 118, *Pl 23: 136*
Purple Emperor Butterfly 53, *Pl 4: 20*
Purple Hairstreak Butterfly 48, *Pl 3: 13*
Puss Moth 97, *Pl 17: 97*
  threat posture 19
PYRALIDAE 150–2
Pyrgus malvae 42, *Pl 1: 4*
Pyronia tithonus 62, *Pl 6: 35*

Quercusia quercus 48, *Pl 3: 13*

Rearing 32–8
Red Admiral Butterfly 53, *Pl 4: 21*
Red Chestnut 120, *Pl 23: 141*
Red Underwing 139, *Pl 29: 178*
Ringlet Butterfly 64, *Pl 6: 38*
Rosy Rustic 136, *Pl 29: 173*
Ruby Tiger 111, *Pl 21: 123*

Sallow Kitten 97, *Pl 17: 98*
Satellite 130, *Pl 26: 159*
Saturnia pavonia 70, *Pl 8: 47*
SATURNIIDAE 70
SATYRIDAE 59–64
Sawfly caterpillars 17
Scarlet Tiger 112, *Pl 21: 124*
Scoliopteryx libatrix 140, *Pl 30: 180*
Segments 15, 17
Selenia dentaria 84, *Pl 12: 74*
Semiothisa wauaria 83, *Pl 12: 71*
Serraca punctinalis 90, *Pl 14: 84*
SESIIDAE 145

Setaceous Hebrew Character 118, *Pl 23: 137*
Short-cloaked Moth 113, *Pl 21: 126*
Shuttle-shaped Dart 114, *Pl 21: 129*
Silk 16, 21
Silver Y Moth 138, *Pl 29: 177*
Silver-washed Fritillary 59, *Pl 5: 30*
Six-spot Burnet 143, *Pl 31: 187*
Sleeving 35
Small Blue Butterfly 49, *Pl 3: 15*
Small Chocolate-tip 101, *Pl 18: 105*
Small Copper Butterfly 49, *Pl 3: 14*
Small Eggar 66, *Pl 7: 41*
Small Ermine Moth 147, *Pl 31: 194*
Small Heath Butterfly 63, *Pl 6: 37*
Small Magpie Moth 151, *Pl 32: 202*
Small Pearl-bordered Fritillary 56, *Pl 5: 26*
Small Skipper Butterfly 41, *Pl 1: 1*
Small Tortoiseshell Butterfly 54, *Pl 4: 23*
Small White Butterfly 45, *Pl 2: 9*
  as pest 26
Smerinthus ocellata 94, *Pl 15: 92*
Snout 140, *Pl 30: 181*
Speckled Wood Butterfly 59, *Pl 6: 31*
SPHINGIDAE 92–6
Sphinx ligustri 93, *Pl 15: 90*
Spilosoma lubricipeda 110, *Pl 20: 120*
Spilosoma luteum 110, *Pl 20: 121*
Spinnerets 16
Spiracles 17
Spruce Carpet 80
Square-spot Rustic 119, *Pl 23: 138*

Stauropus fagi 98, Pl 17: 100
Swallow Prominent 100, Pl 18: 103
Swallow-tail Butterfly 43, Pl 1: 5
Swallow-tailed Moth 85, Pl 12: 75
Sweep net 29
Sword Grass 129, Pl 26: 157
Sycamore 131, Pl 27: 162
Synanthedon salmachus 145, Pl 31: 191

Thera obeliscata 80, Pl 11: 65
Thera variata 80
Theria rupicapraria 91, Pl 14: 87
Thyatira batis 72, Pl 9: 51
THYATIRIDAE 72–4
Thymelicus sylvestris 41, Pl 1: 1
Tinea pellionella 146, Pl 31: 193
TINEIDAE 146
Tineola bisselliella 146, Pl 31: 192
TORTRICIDAE 148–50
Tortrix viridana 150, Pl 32: 200
Trichiura crataegi 65, Pl 7: 40
True Lover's Knot 117, Pl 22: 134
Turnip Moth 114, Pl 21: 128
Twin-spot Carpet 81, Pl 11: 67
Tyria jacobaeae 112, Pl 21: 125

V-moth 83, Pl 12: 71

Vanessa atalanta 53, Pl 4: 21
Vapourer Moth 102, Pl 18: 107

Wall Butterfly 60, Pl 6: 32
Warehouse Moth 151, Pl 32: 203
Warning coloration 20
Wasps, parasitic 22–3
Waved Umber 88, Pl 13: 81
White Admiral Butterfly 52, Pl 4: 19
White Ermine 110, Pl 20: 120
White Satin 105, Pl 19: 112
Willow Beauty 89, Pl 14: 82
Winter Moth 80, Pl 11: 66
   as pest 26

Xanthorhoe fluctuata 77, Pl 10: 61
Xestia agathina 119, Pl 23: 139
Xestia c-nigrum 118, Pl 23: 137
Xestia xanthographa 119, Pl 23: 138
Xylena exsoleta 129, Pl 26: 157

Yellow-tail Moth 105, Pl 19: 111
Yellow Shell 78, Pl 10: 62
Yponomeuta padella 147, Pl 31: 194
YPONOMEUTIDAE 147–8

Zeuzera pyrina 142, Pl 30: 184
Zygaena filipendulae 143, Pl 31: 187
Zygaena trifolii 144, Pl 31: 188
ZYGAENIDAE 143–4

Guide to Famil

SPHINGIDAE 
90

NOTODONTID 
97

LYMANTRIIDAE 
107

ARCTIIDAE 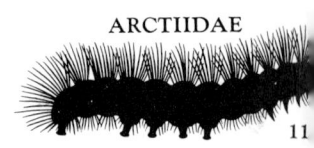
11

NOCTUIDAE 
164

NOCTUIDAE 
165

NOCTUIDAE
178

ZYGAE

Numbers relate to numerical order in text and illustrations